Inside each twisting shell,
   Will a tiny creature dwell,
And when its life is done,
   The shell is ours, what fun!
All that's left to do at last,
   Is walk the beach, find it fast,
   For others with the same desire,
   Scour the sand, and never tire.
The perfect shell, lucky catch,
   Nothing else could ever match,
   Our finding at the tidal stream,
   The collector's hope and
                    sheller's dream.

SANDY BEACHES

Shells Don't Live on the Beach

Live shelled animals cannot survive on dry sand and only a few species such as Coquinas live on wet sand. The rest live under water and only wash ashore after they die. (A few live shells are thrown onto the beach by storms, but perish quickly). Shells found on the beach are the hard external remains of these sea creatures. By the time a shell reaches the beach, the soft body parts of the live animal usually have decayed or been eaten by predators.

Two Kinds of Shells

Notice that the shells on the beach are either single spiral shells, or clam-like shell pairs whose matching sides are hinged together. Many of the paired shells rip apart as they wash onto the beach so that their halves are found separately.

The hinged shells are called bivalves (meaning two shells). The spiral shells are called univalves (one shell). The live animals which inhabit univalves are marine snails.

The two sides of a bivalve shell are joined together with an elastic ligament, interlocking "hinges," and "teeth." Muscles close the shell and the elastic ligament opens the shell when the muscles relax.

How to Use This Book

The shells in this book are not grouped strictly by habitat. Many of the shells which wash up on sandy beaches actually live their lives in offshore waters or shallow bays, but the novice sheller usually encounters them for the first time on the beach. For this reason, these shells have been placed in the Sandy Beaches category. Those shells which are rarely found on sandy beaches have been classified in the habitats where they are most commonly found.

Bivalve

Univalve

Shell Teeth

The Meaning of the Word "Valve"

Many people mistakenly believe that "valves" (as in bivalves or univalves) are the siphon tubes of live shells. But scientifically the word "valve" means the individual shell of a marine snail or the separate halves of a shell pair. In old English, "valve" meant one panel of a folding door, hence the logic of using "valve" to refer to the enclosure of a shell.

Jingle Shells

The wide variations in color make jingle shells fun to collect. Also, the translucent quality of the thin shells makes them very useful in shell craft. A handful of jingle shells will make a jingling sound if shaken, hence their name.

Jingle shells anchor themselves to rocks on the sea bottom or to other shells by means of a byssus *(see page 23 for more details on the byssus)*. The byssal threads pass through a hole in the bottom half of the jingle shell and harden onto the support. If you find a complete pair of jingle shells on the beach, it is an unusual and lucky find. Usually just the top half of the shell breaks loose and washes up on the beach while the bottom half stays anchored in the sea.

The jingle shell is much softer than most other shells. It is easily crushed between the fingers and the shell material cracks into layers of thin, shiny flakes, almost like the mineral mica.

Common Jingle Shell *Anomia simplex** Orbigny, 1842**

* The scientific name of the shell in Latin. The first word is the genus, the second is the species. A shell may have many different common names but only one scientific name.
**The name of the person who first identified the shell in the scientific literature and the date that this information was first published.

Complete shell pairs showing holes for byssal threads

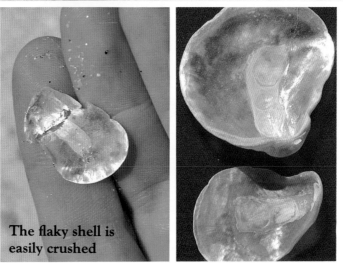

The flaky shell is easily crushed

◁ The jingle has an odd-shaped scar on the inside of the shell where the muscle attaches. It is said to resemble a baby's foot, hence another common name for this shell. The details of this mark vary, but some shells even show a heel and toes.

Moon Snails

Moon snails are so popular that they have acquired many common names including Shark Eye and Cat's Eye. These shells have inspired many paintings such as those by Georgia O'Keefe and poetic writing including that of Anne Morrow Lindbergh, wife of aviator Charles Lindbergh.

Moon snails have big appetites and eat several mollusks every day. They drill holes into bivalves by rasping with their file-like teeth, aided by an acid produced by a special gland. They can drill a hole into a shell in about ten minutes.

Atlantic Moon Shell *Polinices duplicatus* Say, 1822

△ The enormous foot of the live animal can be twice as big as its shell because it is *inflated with water*. When the water is expelled from the flesh, the entire animal can retract into its shell.

"Paul Newman Eyes"
△ Some live moons have strong colors and very intense, blue "eyes." The strong colors tend to fade after the shells have been collected.

◁ The holes in these bivalve shells were most likely drilled by a moon snail.

4

This is a snail shell, round, full
 and glossy as a horse chestnut.

Comfortable and compact, it sits curled up
 like a cat in the hollow of my hand.

Milky and opaque, it has the pinkish bloom
 of the sky on a summer evening,
 ripening to rain.

On its smooth symmetrical face is pencilled
 with precision a perfect spiral, winding
 inward to the pinpoint center
 of the shell, the tiny dark core of the apex,
 the pupil of the eye.

It stares at me — and I stare back.

Now it is the moon, solitary in the sky,
full and round, replete with power. Now it is the
 eye of a cat that brushes noiselessly through
 the long grass at night.

Now it is an island, set in ever widening
circles of waves, alone, self-contained, serene.

Anne Morrow Lindbergh

Other Moons

The Baby's Ear and the Natica are two more types of moon snails found on Florida beaches. The opening of the Baby's Ear shell is so wide that you can look inside all the way to the point of the shell. The Gaudy Natica is distinguished by wavy brown lines and four rows of squarish or arrow-shaped, brown spots.

Baby's Ear Moon *Sinum perspectivum* Say, 1831
• Gaudy Natica *Natica canrena* Linné, 1758

Natica

Baby's Ear

Live Baby's Ear

◁ The live Baby's Ear covers its entire shell and is so large that it cannot pull itself back into its shell like most other mollusks.

Arks

Arks are a type of clam. The Ponderous Ark, which is especially common on Florida beaches, has a black, mossy coating or skin (called a periostracum) on the outside of its shell. This type of coating is common to a number of other shells and may serve as camouflage or protection. However, arks that have been tumbled in the surf will be pure white and frequently appear on beaches in large numbers.

The periostracum is not a living skin. It is actually the outermost layer of the shell and is made of material excreted by the soft parts of the animal (as is the shell).

Ponderous Arks

Transverse Arks

Turkey Wings

Gap for byssus

◁ The fine, soft coloring of the Turkey Wing is said to resemble the feathers of a bird, but it is also called the Zebra Ark. The Turkey Wing lives on rocks just offshore and appears on beaches after storms. The live Turkey Wing attaches to an underwater support with thick byssal threads and there is a noticeable gap between the two halves of the Turkey Wing shell which allows these threads to pass through.

The hinges of the Turkey Wing are much longer than those of most other bivalve shells. All true ark shells have this feature.

Ponderous Ark *Noetia ponderosa* Say, 1822 • Transverse Ark *Anadara transversa* Say, 1822 • Turkey Wing *Arca zebra* Swainson, 1833

The Formation of Foam

Tiny bits of debris from living plants become dissolved in sea water and act as an emulsifier. These organic compounds reduce the surface tension of the water so that bubbles can form, just as soap added to water allows a child to blow bubbles with a toy wand. When the wind whips the water into a froth, foam appears and may create beautiful displays on the beach. Thus foam is usually not caused by pollution but is most common along beaches which are near marshes where organic debris enters the water.

The Sound of the Roaring Sea

How is it possible to hear the sound of the ocean when a large seashell is held to the ear? Many myths have been created to explain the sound including the idea that it is the noise of blood rushing through the arteries in the ear. It may seem less romantic, but a similar sound can be heard from a drinking glass. Like the glass, the shell contains a certain volume of air which slight breezes and faint background noises cause to resonate. Both the shell and the drinking glass act like the sound box of a violin or guitar and amplify this background noise.

If you listen to a shell in a closet with no air currents and lots of clothes to muffle background noise, you will hear very little. Customers in shell shops will frequently pick up a shell and put it to their ear. If they are standing in a quiet corner, they might not hear much. The savvy proprietor will suggest that the disappointed customer move closer to the front door of the shop. There, air currents and noises of the street stir a symphony of sounds from the shell that delight, satisfy, and put the customer in a buying mood.

Religion and Seashells

Orthodox Jewish and Islamic peoples may not see eye to eye in the Middle East, but there is at least one point on which they are in complete agreement: eating shellfish is taboo! Even in Christianity, there is the Old Testament injunction: "Whatsoever hath no fins nor scales in the waters, that shall be an abomination unto you " (Leviticus 11:12). Nevertheless, shells appeared in Christian art and became important religious symbols for early Christians (see Scallop section).

Modern-day Christians generally ignore the dietary restrictions of the Old Testament because of various references in the New Testament including Peter's vision (Acts 10:9-16). The Koran does not specifically ban shellfish but it gives a health warning about them (which probably relates to contamination by a Mediterranean form of red tide). This warning is taken as a prohibition by certain conservative Moslems.

Coquina

Like wildflowers, Coquinas display an infinite variety of colors, and it is a joy to try to choose the prettiest one on the beach. For this reason, and because the dead shells are frequently found in pairs, spread apart like wings, many shellers call them Butterflies. They are also called Wedge Clams.

These fascinating creatures are found alive on Florida beaches just below the sand at the surf line. Coquinas also live along the shores of inlets and bays, but their main habitat is the surf zone: Colonies can contain millions of shells. Run your toe through the wet sand just above the line where the waves are breaking and astonishing numbers may appear. They are most abundant from middle summer through early fall.

The two fleshy "hoses" extending from Coquina shells are siphon tubes. One takes in water and the other expels it after oxygen and suspended particles of food have been extracted. Coquinas can survive by filter-feeding on the plankton that is always present in sea water, and obtain additional nutrients from the particles which are stirred into the water by wave action along the beach.

Coquinas generally stay in the area near the surf line where the sand is constantly wet. In order to remain in this zone, Coquinas must migrate up and down the beach daily as the tide rises and falls. It is thought that the Coquinas on the beach regulate their movements by sensing the vibration of the waves pounding on the sand. When they sense the waves of an incoming tide approaching, Coquinas emerge from the sand and allow the waves to push them shoreward. After being pushed up the beach, the Coquinas dig in to prevent being washed back down.

If a wave pushes the Coquinas too far, they can sense that the sand is too dry and will allow the next waves to drag them back down the beach. The watery soup of the wet sand allows them to burrow in and out quickly. In this manner they manage to stay in the area where the sand is wet through all the tidal changes.

Coquina *Donax variabilis* Say, 1822

△ Beachcombers are always amazed at the speed at which a coquina can bury itself. The coquina will stand straight up and use its large foot to wriggle into the sand in a flash. This mass performance has been called the "dance of the coquinas."

Coquina Broth Recipe

Collect several quarts of coquinas. Make sure they are all alive and that there are no other types of shells mixed in which might not be live and fresh. Rinse well to remove any algae or debris. Cover in a pot with about 1/2" of water. Add a little pepper, a little butter, and a couple of tablespoons of sherry. Boil for 5 minutes. Strain out the shells and serve the broth.

Some people use toothpicks to remove the meat from the shells, but this is tedious and not very rewarding. Coquinas are really best for broth, not chowder, and the broth can be used as a tasty soup stock or base for other dishes.

Whelks

The Lightning Whelk is the only Florida shell which is regularly left-spiraling. Among other species, left-spiraling shells are rarities, freak shells that spiral contrary to the vast majority of their kind. The scientific name, *Busycon contrarium*, celebrates the way the Lightning Whelk grows contrary to most other shells.

Lightning Whelks eat clams and other bivalves. Rather than drilling a hole in clam shells like many other marine snails, the Lightning Whelk uses its foot and the edge of its shell to pry open clams and then uses its proboscis (a tongue-like tube with teeth in the end) to feed on the meat.

Lightning Whelk *Busycon contrarium* Conrad, 1840

△ Baby Lightning Whelks are perfectly formed miniature versions of the mature shells.

△ Note the large size of this Lightning Whelk. The live animal is quite dark in color.

Old Folks Home for Shells

The aging process often changes the color of shells. Note the almost pure white color of large Lightning Whelks compared to the brilliant streaked color of the young shells. Some Horse Conchs go through a similar change of color (*see page 54*). But there are some shells such as the tulips which do not change in color as they grow larger and older. Very large whelks may be over 10 years old. Also, the largest whelks are undoubtedly females since the females are generally much larger than the males. Large whelk shells are popular as decorations in Florida gardens.

△ The Lightning Whelk gets its common name from streaks on the shell which resemble flashes of lightning.

Knobbed Whelk

△ In Florida, Knobbed Whelks are found only along the east coast from Cape Canaveral northward, but they are very common in that part of the state. Their range extends as far north as Cape Cod.

Knobbed Whelk *Busycon carica* Gmelin, 1791

◁ A Lightning Whelk with its feeding tube extended. Known as a *radular proboscis*, this remarkably long organ is common to many marine snails.

Hindu priest blowing shell horn

A New Wrinkle in the Old Shell Game

In India there is a certain species of shell (called a Chank) which coils to the right, as most shells do. However, very rare specimens of this species coil to the left. These rare shells have religious significance to both Hindus and Buddhists (and are too valuable to make into horns). Some enterprising Indian businessmen learned that Florida's Lightning Whelk naturally coils to the left and transported this relatively cheap shell to India to be sold to the faithful for easy profits.

The horn shown in this photo was created by grinding the shell against a rock to remove the tip of its spire. It is played like a trumpet by blowing into it with vibrating lips. The sound is impressively loud.

Left-handed, Right-handed?

Hold a spiral shell with its spire pointing up and its channel pointing down. If you can insert your right hand into the opening, it is right-handed. If it invites you to insert your left hand, it is left-handed. Except for Lightning Whelks (which are mostly left-handed), a left-handed shell is a freak of nature and a rare, lucky find for a sheller in Florida. These wrong-coiling shells are called "sinistral specimens" by collectors.

Left-handed

Heads and Tails

The pointed end of a spiral shell (the apex) is actually the tail of the animal. The animal's head emerges from the shell at the open, channeled end (siphonal canal).

Cockles

Van Hyning's

Yellow

Morton's Egg

Common Egg

Prickly

Cockle shells are distinguished by their heart shape. The Prickly Cockle has the sharpest spines along the ribs. The Egg Cockles have smooth shells and a more oval shape. Van Hyning's Cockle is the largest of Florida's cockles and features dark-red color. The Yellow Cockle has prominent ribs.

Note the beautiful pink color on the inside of the Van Hyning's Cockle. These shells are large enough to use for ashtrays and small containers.

Europeans and Asians love to eat cockles. Cockles are immortalized in the Irish folk song, "Sweet Molly Malone."

In Dublin's fair city,
Where the girls are so pretty,
There once lived a maiden
named Molly Malone.
She wheeled her wheelbarrow,
Through streets wide and narrow,
Crying "Cockles and mussels,
alive, alive, oh."

△ Live cockle showing siphon-tubes

Prickly Cockle *Trachycardium egmontianum* Shuttleworth, 1856 • Yellow Cockle *Trachycardium muricatum* Linné, 1758 • Common Egg Cockle *Laevicardium laevigatum* Linné, 1758 • Van Hyning's Cockle *Dinocardium robustum vanhyningi* Clench and L. C. Smith, 1944 • Morton's Egg Cockle *Laevicardium mortoni* Conrad, 1830

Valentines from the Sea

Cockles are sometimes known as Heart Clams because they have a heart-like shape when they are viewed from the side.

Safe Clamming

Bacteria in polluted waters usually attach themselves to small particles rather than floating freely. These particles are taken in as part of the food filtering process of live clams. The bacteria may accumulate and reach dangerous levels in shellfish which are filter feeders such as cockles, clams and scallops. For this reason it is best to gather shellfish in areas which have been tested and proven safe by the Florida Marine Patrol. Call the patrol or the county public health service for current information.

Why Sanibel is So Special

There is a wide plateau of relatively shallow water and sandy bottom adjoining Sanibel on the south side of the island. Miles offshore the water is less than 40 feet deep. (By comparison, in the Florida Keys the depth could be as much as 300 to 800 feet.) This gradual slope of the Gulf bottom acts like a ramp and allows large numbers of shells to roll onto the beach, especially when driven by storms from the northwest. These storms are common in December and January when cold fronts pass through the state. The gentle slope also assures that many shells arrive in undamaged condition.

Sanibel Island

The alignment of Sanibel is mostly East/West whereas the other barrier islands hug the coastline in a North/South direction. By sticking out into the Gulf, Sanibel becomes a giant scoop for shells pushed northward by prevailing coastal winds and currents which are from the South.

For these reasons, Sanibel is considered one of the best places in the world for beach shelling, although in Florida, the Keys are a better place to snorkel or dive for live shells.

Compared to the East Coast of Florida, the West Coast beaches are far richer in shells. The reason is that the drop-off into deep water is much steeper on the East Coast, creating a wall-like barrier which prevents many shells from reaching the beach. The abrupt change of depth also creates pounding surf which pulverizes the shells. On the East Coast it is not unusual for the water to be 70 feet deep a mere hundred yards offshore. However, the shells that do reach the East Coast beaches are more likely to be deep-water species.

Naturally, the folks on Sanibel make the most of their good fortune and try to build their reputation as the shell capital of the world. You will even notice a lot of streets that have been given shell names, such as Donax Street and Periwinkle Way.

Best Beach Shelling

The best beach shelling is often at the ends of islands facing the passes and channels between islands where the strong currents and tidal flows move many shells.

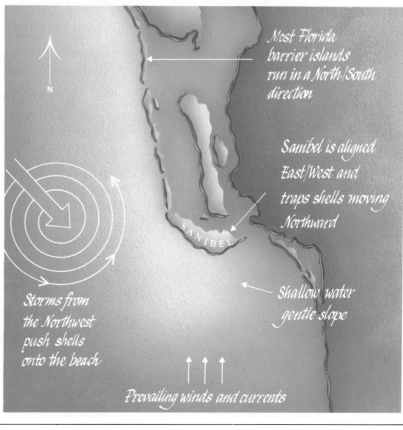

Most Florida barrier islands run in a North/South direction

Sanibel is aligned East/West and traps shells moving Northward

Shallow water gentle slope

Storms from the Northwest push shells onto the beach

Prevailing winds and currents

Scallops

Bay

Sentis

Rough

Calico

Lemon Pecten

Scallops are among the most beautiful of shells because of their fantastic variety of color. The scallop is recognized by people all over the world as the symbol of the giant Shell Oil Company, a firm which began business as an importer of seashell curios for wealthy collectors. Even today Shell Oil tankers bear the names of seashells in homage to the company's humble origin.

Throughout the Middle Ages, the scallop was an important religious symbol. Land routes to the Holy Land had been cut off by infidels, so as an alternative, the pious made pilgrimages to the grave of St. James on the coast of Spain. They began bringing back scallop shells from the Spanish beach, often on necklace chains, as symbols that they had completed their journey. The scallop was also adopted as the symbol of the Christian crusaders who attempted to recover the Holy Land from the Muslims. It appears in the coats of arms of many British families (including Winston Churchill's).

"The Scallop shows a coat of arms,
that, of the bearer's line,
Someone in former times,
Hath been to Santiago's shrine."

Thomas Fullerton

The Bay Scallop is the most desirable of the edible scallops, but because of availability, the type usually served in restaurants is the smaller Calico Scallop. Bay Scallops do live in bays, but many of their shells wash up on the open beaches. The upper halves of these scallop shells are often darker (perhaps to provide camouflage) while the lower valves may be very light. The Rough Scallop is hard to find because it tends to be covered with algae and sponges, thus camouflaging itself. The shell has spines along each rib, thus giving it the common name of Rough Scallop. Its yellow form is called the Lemon Pecten and is highly desired by collectors. The small Sentis Scallop comes in many different, often bright colors. It is found attached by its byssus to the bottom of rocks, mainly in the Keys. It is flatter than most of the other Scallops. The Calico is the scallop most commonly eaten in Florida. It is harvested in the deep waters off Cape Canaveral. Its color is rather varied and usually includes some mottling. The Calico is the scallop most often found on Florida beaches.

Bay Scallop *Argopecten irradians* Lamarck, 1819
Rough Scallop *Aequipecten muscosus* Wood, 1828
• Sentis Scallop *Chlamys sentis* Reeve, 1853
• Calico Scallop *Argopecten gibbus* Linné, 1758

Do Scallops Migrate?

Most scallops are free swimming and move through the water by opening their shells and clapping them shut, squeezing jets of water from the shells. This ability to swim sometimes enables them to escape starfish, which are their most dangerous predators. Although scallops can propel themselves for short distances, they do not migrate as commonly believed. Because of the explosive growth of young scallops and the dramatic die-offs of old scallops, scallop populations can change very suddenly. For this reason it sometimes appears that scallops are moving in and out of bays in large numbers.

Live Calico Scallop

All Those Eyes

The scallop has rows of eyes, each eye having its own lens and retina. The lens collects and focuses light. However, the scallop lacks a center of vision in its brain and probably cannot form an image. So what good are all these beautiful eyes? The scallop can probably distinguish light from dark and this may help it move from sandy areas to grass flats. It may be able to sense movement and this would aid in avoiding predators. A diver casting a shadow over a scallop will cause it to snap its shell closed.

The Great Scallop Scam

In the bad old days it was common for restaurants to economize by cutting plugs from the wings of sting rays or from shark meat and substituting them for the more expensive scallops. Since the rise of the scallop industry on Florida's East Coast in the 1970's, this practice is unnecessary and rare. The best way to tell the difference is to break the "scallop" meat apart. A true scallop will break apart easily and all the fibers will be running in the same direction.

"Black Scallops"

Black or dark-gray scallop shells are common on the beaches. But these dark scallops are not a separate species. The shells shown in the photo are all Calico Scallops. In the dark shells, a chemical reaction has occurred in which some of the calcium carbonate in the shells has been replaced with iron sulfide. This process is similar to the fossilization of bone, but can occur within a few years time when scallops are buried in the offshore muck and no oxygen is present. Other shells, including venus clams and arks, are subject to this darkening process, but scallops seem to be the most vulnerable.

Note: four more scallop species are discussed in the chapter on deep-water shells.

Kitten's Paw

The Kitten's Paw or Cat's Paw is a small, scallop-like shell which is not in the scallop family. Strong ribs help create the likeness of a paw.

The Kitten's Paw usually attaches itself to other shells or rocks with a very powerful marine glue which the animal manufactures. For this reason, it is usually only one half of the shell that tears loose and washes up on the beach. It is difficult to find a complete pair unless you find both shells still attached to their base.

The Kitten's Paw is a small shell, less than an inch in length. It is very common on the beaches of Florida and should not be confused with the spectacular Lion's Paw which is much larger and lives in deep water *(see page 85)*.

Atlantic Kitten's Paw *Plicatula gibbosa* Lamarck, 1801

Fig Shell

This shell is called the Common Fig, the Paper Fig and also the Atlantic Fig. It is considered uncommon except on the West Coast of Florida. The very thin shell is easily damaged and rarely found in good condition on the beach. It has no trapdoor (operculum).

Fig shells are often confused with Pear Whelks. One difference is the texture. The surface of the fig shell is decorated with a fine criss-cross pattern. A rare fig shell may have a bluish color on the inside.

Common Fig Shell *Ficus communis* Röding, 1798

All Those Little Tapered Shells

There are several families of small, tapered shells that are tricky to identify because they are about the same size and shape. Here they are, all together, so that the differences are more obvious.

▷ Horn Shells are so named because of the horn-like shape of the shell opening or aperture.

Auger *Terebra* spp. • Horn Shell *Cerithidea* spp. • Turrets *Turritella* spp. • Cerith *Cerithium* spp. • Worm Shell *Vermicularia* spp.

Wentletrap

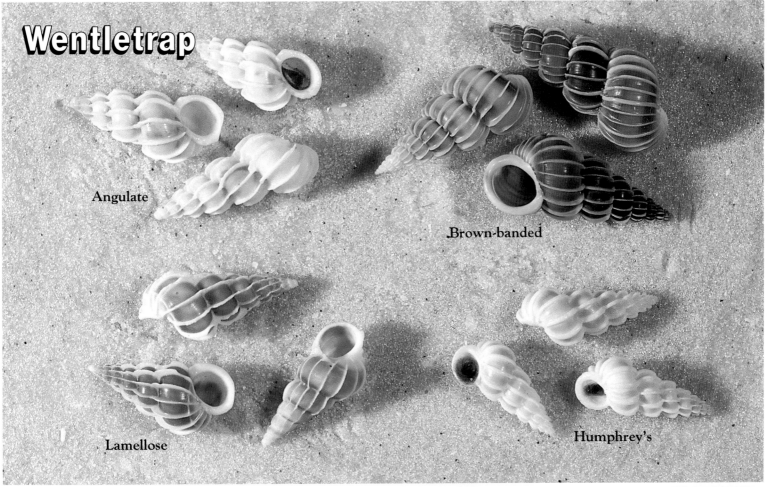

Angulate

Brown-banded

Lamellose

Humphrey's

Because of the pattern of ribs on the outside of their shells, wentletraps are called Staircase Shells. In fact, the name "wentletrap" comes from the German word "wendeltreppe" meaning "spiral staircase." Angulate Wentletraps are quite common on Florida beaches, especially on Sanibel Island where they can be found in the debris left at the high-tide line. The beach around the lighthouse is a good place to search.

- Humphrey's Wentletrap *Epitonium humphreysi* Kiener
- Angulate Wentletrap *Epitonium angulatum* Say, 1830
- Lamellose Wentletrap *Epitonium lamellosum* Lamarck, 1822
- Brown-banded Wentletrap *Epitonium rupicola* Kurtz, 1860

Live Humphrey's Wentletrap

Nutmeg

These shells are said to resemble the shape of a nutmeg seed. They are common in 20 to 30 feet of water, but not always easy to find on the beach. Nutmegs are beaded on their outer surface and usually brownish in color, but albino specimens can be found. There are many attractive nutmeg species worldwide, and their variety has made the nutmegs popular with collectors.

Common Nutmeg *Cancellaria reticulata* Linné, 1767

Break Line

Folds

◁ Note the two folds inside the shell opening. The upper fold is usually larger. Also note the break lines which are visible on the outside of most nutmegs. These lines are scars which mark places where the shell has been broken (perhaps by a hungry crab). The nutmeg repairs the damage, but a break line remains.

Bubble Shells

Like the moon snail, the live animal is many times the size of its shell. These small shells belong in the same subclass as sea hares and nudibranchs, both marine snails with little or no shell. Bubble Shells likewise have reduced shells which are rather thin and fragile like bubbles and can be broken between your fingers. Like the cowries, the shell opening extends the entire length of the shell. The live animal is generally colorful. Live bubbles are usually found at night when they are most active.

Common Atlantic Bubble *Bulla striata* Bruguiere, 1792

Buttercup

The Buttercup Lucina is almost circular and quite concave. Sometimes pairs of shells can be found still joined together, although the individual shells are much more numerous on the beach. This shell is very popular in shellcraft because of its deep yellow color inside, which is reminiscent of the buttercup flower.

◁ The Buttercup Lucina is found on sand in a wide range of depths from very shallow water down to 300 feet.

Buttercup Lucina *Anodontia alba* Link, 1807

Ivory Tusks

These shells are seldom more than an inch in length and can easily go unnoticed. An interesting feature is that the shells are open on both ends. This makes the shells easy to string for necklaces. Because of their pointed, tooth-like shape, tusk shells are popular for jewelry in many parts of the world.

Tusk shells are usually found in batches among the piles of shells on the beach, perhaps several dozen in one spot and no others nearby. Many species of shells are found clustered in this manner because they live close together in areas where conditions are good for feeding or breeding and are often washed up on the beach in a group.

Tusk shells are neither univalves nor bivalves but belong to a rather primitive class by themselves. They feed on microscopic organisms.

The live animal lives in an upright position with its larger end anchored in the sand.

△ Tusk shells with Coquinas for size comparison.

Ivory Tusk *Dentalium eboreum* Conrad, 1846

18

The dead shells of this species are sometimes common on West Coast beaches. They are usually heavily worn and damaged when found on the beach because they do not live at the surf line. They live in slightly deeper water (5 to 25 feet deep) and only reach the beach after they die and are tumbled ashore by the action of waves and tides. However, they do come into shallow water in the spring to breed and lay eggs. If you find a good specimen, notice the brown color outside contrasted with the variable colors inside, including the rare purple.

Some say this shell was named a fighting conch because the points on the shell resemble the spikes worn by ancient gladiators. Others point to the animal's quick movements in which it uses its pointed "foot" to propel itself. Most mollusks, if picked up, will pull back into their shells and close their doors in a turtle-like, defensive maneuver. But not the Florida Fighting Conch. If picked up by a collector, it will come out of its shell and make vigorous efforts to escape by thrashing about, trying to dislodge itself.

Florida Fighting Conchs are algae eaters and usually graze in large colonies like herds of cattle in a pasture. Despite their ferocious-sounding common name, if bothered by other mollusks, they only try to escape.

Florida Fighting Conch *Strombus alatus* Gmelin, 1791

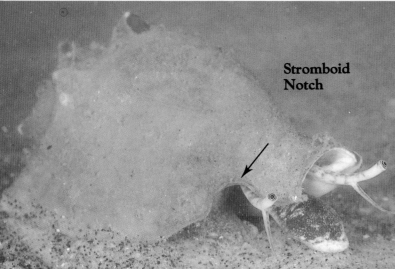

△ The eyes of the fighting conch extend outward from the shell on long stalks. Note one eye extending through the "stromboid notch." This notch is common to all true conch shells (*see photo at bottom of page 70*).

Stromboid Notch

Live Fighting Conch

△ Life Cycle: Note that the immature shell is a bit different from the adult and is often mistaken for a cone shell.

On the East Coast of Florida, at Lake Worth Inlet, there is an interesting colony of very large, dark-brown to black, color forms of the Fighting Conch.

Slipper Shells

Slipper shells are among the most common Florida shells found on open beaches. It is easy to imagine these shells as slippers, but they also resemble small boats with a little "seat" inside. A slipper shell which is well shaped and balanced can be floated on calm waters, a game which is always fun for children at the beach. It is not surprising that they are called Boat Shells, Canoes, and Quarterdecks.

Young slipper shells are quite mobile, but the older shells live in groups, which makes mating more convenient.

Atlantic Slipper Shells

Slipper shells stacked together

Spiny Slipper Shells

Mature slipper shells stack up on top of each other (sometimes eight high) and live attached together in chains. These stacks of shells are usually arranged with the males at the top. During mating, the sex organs of the male shells on top will protrude downward into the female shells lower in the stack.

Fantastically, when the females at the bottom of the stack die and no longer excrete sex hormones, the males stacked just above them change their sex to female, so there are always females at the bottom of the pile.

Slipper shells are gastropods, single-shelled marine snails, and they are among the few marine snails which are not coiled in a spiral shape.

Atlantic Slipper Shell *Crepidula fornicata Linné,*
1758 • Spiny Slipper Shell *Crepidula aculeata Gmelin, 1791*

Slipper Shells
on Horse Conch

◁ Slipper shells frequently attach themselves to other shells. They follow and conform to the shapes of their host shells as they grow (*see also photo at top of page 65*).

Watch the Birdie

Since so many beach creatures are hidden just under the sand, they are not often visible to the human eye. However, their presence is betrayed by flocks of feeding shorebirds. By digging where the birds are digging, it is easier to locate many of the live creatures of the surf zone.

Olives

Lettered Olives

The markings on this shell resemble hand lettering (written by someone with very bad penmanship).

Olives are found in the sand at the surf zone and just below it. They crawl through the sand searching for small clams such as coquinas. They don't have good vision and must rely on their sense of smell to find food.

When found alive, olive shells are very glossy because the fleshy mantles of the live animals extend around the outside of the shells and protect them. Many visitors to shell shops think that these shells have been artificially polished, but it is a natural process (*see page 79*). When found dead on the beach, olives are usually dull after much tumbling in the sand and surf.

Lettered Olive *Oliva sayana* Ravenel, 1834

Live Olive

Golden Olive

Lettered Olive

◁ Olives stranded at low tide push their shells through the mud looking for water. They leave trails in the sand which alert shellers to their locations.

△ The Golden Olive is simply a yellow color form of the Lettered Olive, but it is difficult to find and is eagerly sought by collectors.

Tellins

Speckled

Rose Petal

Sunrise

Sunrise Tellins are similar in color to Coquinas, but are three to four times larger. Rose Petal Tellins display a beautiful range of colors, from cream color to deep pink.

Rose Petal Tellin *Tellina lineata* Turton, 1819 • Speckled Tellin *Tellina listeri* Röding, 1798 • Sunrise Tellin *Tellina radiata* Linné, 1758

Modulus

▷ The Modulus is also known as the Button Shell because of its small size (less than 1/2"). Although the dead shells are found on beaches, the live shells are common in shallow bays.

Atlantic Modulus *Modulus modulus* Linné, 1758

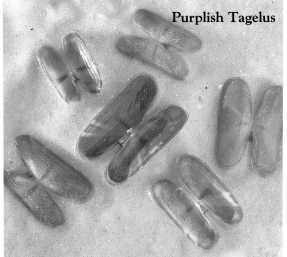

Purplish Tagelus

Tinted Cantharus

◁ This fragile shell lives in the sand between the high tide and low tide lines. It is usually found as a hinged pair and generally has a purplish tint.

Purplish Tagelus *Tagelus divisus* Spengler, 1794

◁ Cantharus are members of the whelk family and are known to cause damage to oyster beds by feeding on small oysters. These small shells are usually found on rocks but sometimes wash up on beaches. Their color can be orange, brown, purple, or even black.

Tinted Cantharus *Pisania tincta* Conrad, 1846

22

Pen Shells

Many pen shells wash up on Florida beaches after storms because they live in shallow waters near the shore. Pen shells are distinguished by their brittle thinness and the sharp spines on their outer surfaces. Because of their shape, they are also called Fan Shells. A variety of sea creatures frequently attach themselves to the outer surface of dead pen shells, including oysters, barnacles, slipper shells, and chitons.

The Fascinating Byssus

Pen shells attach themselves to rocks or to the sandy bottom with an anchor chain called a byssus (pronounced "biss-us"). The byssus is composed of strands of flexible material which attach to tiny shell fragments. These pieces of shell act like little anchors when buried in the sandy bottom. Other species of shells which have this attaching ability include the mussels, jingle shells, ark shells, file clams, jewel boxes, turkey wings, and certain oysters and scallops.

Stiff Pen Shell *Atrina rigida* Lightfoot, 1786 • Saw-toothed Pen Shell *Atrina serrata* Sowerby, 1825

Saw-toothed Pen Shell

Stiff Pen Shell

Like Gold Wrapped in Rags

The Stiff Pen Shell may be the ugliest shell on the beach, but inside there is a treasure, a beautiful iridescent layer. If you find an iridescent scrap of shell on the beach, most likely it was once part of a pen shell that has been broken and polished by the waves. Pen shells should not be confused with the very iridescent abalone shells (which are not found in Florida).

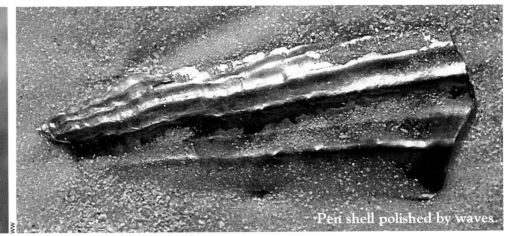

Byssus with shell fragments as anchors

Pen shell polished by waves.

Pen Shells and the Golden Fleece

If washed and combed, the threads of the pen shell byssus become fine and silky. The byssus of a particularly large Mediterranean pen shell was processed by ancient peoples (and sometimes mixed with real silk) to produce small items of clothing such as gloves and socks for a very rich clientele. This material may have been the "golden fleece" sought by the ancient Greeks. Modern day shell crafters in Florida sometimes collect and process the byssus of the pen shell to make hair for figurines which are constructed entirely of sea-related materials (*see page 109*).

Mother-of-Pearl △

The beautiful iridescent finish of some shells is created as the live animal lays down many very thin layers of a translucent material called nacre. This mother-of-pearl material is actually the same substance found in pearls. In some shells such as the Stiff Pen Shell, it is very colorful. Many fresh water shells display beautiful mother-of-pearl (*see page 110*).

The Saw-toothed Pen Shell lacks the sharp spines of the Stiff Pen Shell, and while perhaps more attractive on the exterior, the iridescent lining inside lacks the multiple colors found in the Stiff Pen Shell.

23

Clams

the most common shell on the beach?

Cross-barred Venus Clam

△ Along with Coquinas, these small clams are probably the most common of the dead shells on Florida beaches. They are found as single shells and complete pairs. The cross-hatched pattern on the outside and the purple color inside make identification easy. These shells are often found piled up on the beach in great mounds at the high tide line and around jetties and obstructions where shells and debris tend to collect.

Cross-barred Venus Clam *Chione cancellata* Linné, 1767

Sunray Venus Clam

The live Sunray lives in the sand that is sometimes exposed at low tide. This shell is very common in Florida. A number of years ago these clams were gathered commercially for the seafood market. Most of this commercial activity was in the area of Port St. Joe in the Florida panhandle. Sunrays are a favorite food of gulls and terns as well as humans.

Sunray Venus Clam *Macrocallista nimbosa* Lightfoot, 1786

◁ Many shells found on the beach are defaced by grooves which look like they were cut with a routing tool. This kind of damage may be caused by boring sponges or marine worms seeking a foothold, or from the efforts of a marine snail predator, such as a whelk, attempting to bore into the shell with its radular teeth.

Sunray Venus Clam

Harvesting Sunrays for Dinner

Calico Clam

The Calico Clam is not as common as the Sunray Venus Clam, but it is beautiful and also very delicious.

Calico Clam *Macrocallista maculata* Linné, 1758

Surf Clam

This is one of the most important of the clams harvested for food. (The other important food clams are the Quahogs and the Sunrays.) It is easy to recognize because of the tan coating (periostracum) on its exterior. These clams are harvested commercially by dredging the ocean floor.

Atlantic Surf Clam *Spisula solidissima similis* Say, 1822

Calico Clam

Atlantic Surf Clam

Quahog

Dosinia

Quahog

These clams live in the shallow waters of bays. The small young Quahogs (pronounced "coh-hogs") are more tender and desirable but are harder to find because they grow to adult size very quickly and because the small quahogs bury themselves more deeply. The adults are also edible, but rather tough. Their meat is used by fishermen for bait. The empty shells of large specimens are popular for ashtrays.

Southern Quahog *Mercenaria campechiensis* Gmelin, 1791

Dosinia

These round, flat shells are particularly common on the central West Coast of Florida. The Disk Dosinia has very fine circular ridges like the grooves of a stereo record.

Disk Dosinia *Dosinia discus* Reeve, 1850

Duck Clam

Also known as a Sailor's Ear, this thin, delicate shell has ridges which are more widely spaced than the dosinia and a rather triangular shape.

Channeled Duck Clam *Raeta plicatella* Lamarck, 1818

Channeled Duck Clam

Murex

Murex shells are known world-wide for their spectacular frills and spines. The shell opening (aperture) is unusually round and located at the middle of the shell. The Lace Murex is most common in Florida. With its delicate frills at the tips of its spines, it is a favorite with shellers. The color of the Lace Murex can range all the way from black to white, with the intermediate brown tones most common. The small Hexagonal Murex is found on reefs in the Florida Keys. The Giant Murex is the largest member of the Murex family in Florida. It can be found in shallow water in northern Florida, but it is more often trapped by shrimpers trawling in deep water.

Apple Murex *Phyllonotus pomum* Gmelin, 1791 • Florida Lace Murex *Chicoreus florifer dilectus* Adams, 1855 • Giant Eastern Murex *Hexaplex fulvescens* Sowerby, 1834 • Hexagonal Murex *Muricopsis oxytatus* M. Smith, 1938 • Rose Murex *Murex rubidus* F.C. Baker, 1897

Giant Eastern

Apple

Lace

Hexagonal

Rose

Murex shells were famous in ancient history as the source of a rare fabric dye used by the Romans for their "royal purple" robes. Thousands of shells were required to extract just a tiny amount of the dye by drying and boiling the soft body parts of the murex animals. Thus, only the very rich and powerful could afford to wear this color. Yet Antony and Cleopatra were extravagant enough to dye all the sails of their ship purple for the battle of Actium. In nature the dye fluid is used by the murex to slowly anesthetize prey such as clams. The foul smelling liquid is actually yellowish in color but turns dark blue when exposed to sunlight.

"Who hath not heard how Tyrian shells
Enclose the blue, that dye of dyes,
Whereof one drop worked miracles,
And colored, like Astarte's eyes,
Raw silk the merchant sells?"

Robert Browning

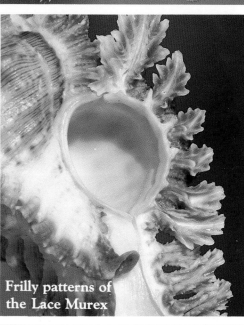

Frilly patterns of the Lace Murex

Cardita Shells

Many halves of the shell pairs are found on beaches, especially on the West Coast of Florida. The complete shells are more often found in bays. They are a favorite food of the larger marine snails such as conchs, tulips and whelks. They are easily recognized by their prominent ribs.

Cardita shells are not much favored by collectors but are popular in shellcraft. They are used to form the feet of many small novelty critters.

Broad-ribbed Cardita *Carditamera floridana* Conrad, 1838

OTHER BEACH LIFE

Jellyfish

There are many kinds of jellyfish in Florida waters. Most of them have stinging cells and can give your skin a temporary burning sensation if you happen to bump into one while you are in the water, although the sting is not as dangerous as that of the Portuguese Man-of-war. Ammonia and meat tenderizer both help relieve the pain of jellyfish stings.

The Moon Jellyfish, one of the common Florida species, is a beautiful, graceful creature when it is seen swimming. It is, incidentally, a very strong swimmer (for a jellyfish).

The Moon Jellyfish is not dangerous to man, but it can sting and irritate the skin and some people are especially sensitive. It is frequently found on the beach and the male and female can be identified (the male sex glands visible inside are pinkish, the female's brown).

Sea turtles love to eat Moon Jellyfish and will sometimes mistake a floating plastic bag for their favorite delicacy. This mistake can be fatal to a turtle, so beachcombers are cautioned to not discard plastic bags into the water. Environmentalists recently protested a school project involving the release of thousands of helium balloons. They were concerned that if the partially deflated balloons fell into the ocean, sea turtles might mistake them for Moon Jellyfish.

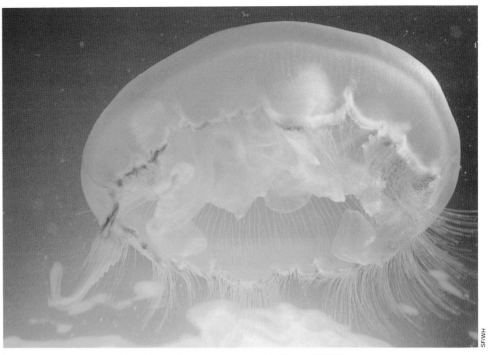

SFWH

△ The Moon Jellyfish has a distinctive row of short tentacles around the bottom edge of its bell and can be two feet wide.

◁ Sea Nettles feature reddish, radial markings. These jellies can really sting.

Upsidedown Jellyfish (Keys)

Cannonball Jellyfish

Comb Jelly glowing in the dark. In daylight they are quite transparent

Comb Jelly

◁ This jelly-like animal is most frequently found stranded on the beach or swimming in the surf. Comb Jellies are small (2 to 5 inches), fragile, and almost totally transparent. It is fascinating to watch the play of iridescent light along their 8 "comb rows." The "teeth" of the combs are actually very fine hairs or cilia that beat rhythmically when the animals swim. This animal (it is not a true jellyfish) does not sting. Some Comb Jellies are luminous, and sometimes one that has washed ashore can be seen glowing on the beach at night. Their glow in the water is familiar to night fishermen. The scientific name of this creature is *Ctenophore*, pronounced "10-Oh-4" and also "Teen-Oh-4."

Sand Dollar

Arrowhead
Sand Dollar

"Doves"

Flat Sand Dollar

The sand dollar consists mostly of hard shell (called a test) with very little live tissue. Yet, this creature manages to capture more than its share of attention from beach-goers. Not only is the sand dollar interesting and beautiful, but the holes and markings in the hard test are seen as religious symbols by many.

The five small holes around the center on the top side of the sand dollar are "gonapores" through which the live animal emits eggs or sperm into the water. Notice the flower pattern on the top side of the sand dollar. This pattern is created by a series of very small slits, clearly visible on a dried specimen. When the animal is alive, tiny appendages are extended through the small openings to exchange gases with the water. This is how the animal breathes.

The five large slots are called lunules. One function of these openings is to speed burrowing by allowing sand to be passed up through the sand dollar as the creature digs its way downward. They also allow food particles that settle on top of the sand dollar to be passed downward to the mouth.

The teeth are arranged in a star shape just inside the mouth opening on the bottom. This arrangement of chewing equipment is called Aristotle's lantern.

Gonapores

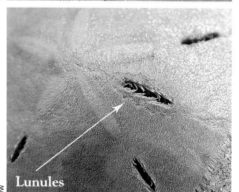

Lunules

Aristotle's Lantern

Legend of the Sand Dollar

The sand dollar has been called the Holy Ghost Shell because of the religious symbolism of its markings. It is said to tell the story of the birth, crucifixion and resurrection of Christ.

The outline etched into the top resembles an Easter Lily. At the center is the five-pointed star of Bethlehem. The five oval holes represent the wounds Christ suffered on the cross (the four small holes are nails in the hands and feet, and the larger hole the wound from the spear).

On the bottom of the sand dollar is the outline of the Christmas flower, the Poinsettia. If broken open, five of the sand dollar's teeth are seen arranged in a star shape. Individually, these little pieces resemble white doves in flight which some say are the five angels that sang to the shepherds on Christmas morning.

Harmless yellow dye stains fingers

Live sand dollars leave a harmless, yellow stain on the hands of the collector. This dye has been utilized to color fabrics in some parts of the world. While holding the live sand dollar, watch its tiny spines moving together in wave-like patterns. The spines on the bottom allow it to move about (although slowly) and to dig into the sand.

Live sand dollars are quite sturdy, but dead sand dollars found on the beach are very soft and easily broken. The sturdy specimens found in shell shops have been processed by leaving them outdoors for many weeks and washing them every few days. If cured in this manner, they become strong enough to be collectible.

The sand dollar is found alive in shallow water mostly during summer and fall. It is easy to locate when wading because it will be only partially covered with sand. You will notice a coarse-feeling, flat object under your feet and you can reach down and pick it up. If the water is shallow enough, you can see its tell-tale pattern in the sand.

Radial Symmetry

Although the sand dollar may not appear very similar to the sea urchin or the starfish at first glance, they are all echinoderms, creatures which are designed like the spokes of a wheel, a shape scientists call radial symmetry.

Bottom is covered with small, moveable spines

Tell-tale pattern of buried sand dollar

Test versus Shell

The hard skeletons of sand dollars (as well as sea urchins) are called tests rather than shells. Tests are made of the same material as shells but in the live animals are usually covered with an external, living, skin-like layer.

29

Egg Cases

Egg Case of Lightning Whelk

Many sea creatures lay their eggs in special containers, and in the springtime these containers frequently wash up on beaches where they are a puzzlement to novice beachcombers. What to make of the very long and very large egg case of the whelk? Many people guess that it is the backbone and skeleton of a snake. Few would guess that a mollusk would produce anything so large.

Each pocket of the Lightning Whelk egg case contains between 20 and 100 tiny whelks. If the egg case is fertile, the beachcomber can cut open a pocket and examine the baby whelks.

Baby Lightning Whelks

△ Banded Tulip laying eggs in its unique wavy case. The egg case of the Banded Tulip is often found attached to the egg case of the Lightning Whelk.

Egg case of Crown Conch

△ These tiny whelks begin to eat each other like cannibals as soon as they emerge from the case. Most of the eggs in each capsule serve as food for the usually single individual that will emerge. In northern waters, the young remain in the capsules longer than in Forida.

These tiny shells are popular with shell crafters for miniatures.

Moon Snail Collar

"Devil's Purse" of Skate

The sand collar (or sea collar) produced by the moon snail is formed when eggs are laid and serves to protect the eggs. The eggs are mixed with sand and a gelatin-like substance which binds the mixture together, although the finished collar remains quite fragile. Fragments of this collar are frequently found on Florida beaches and bay shores. Perfect collars are easily found in the quiet waters of bays.

The collars are very flexible (and fragile) when wet, but they can be picked up and collected if great care is used. Collectors sometimes soak sand collars in a mixture of alcohol and glycerin or spray them with lacquer, so that the fragile structures will hold together for display.

The egg case of the clearnose skate (sometimes called a mermaid's purse or devil's purse) is also fascinating. It is made of a material which is similar to that of fingernails. Unopened, it looks like it might be a seed pod with hooks or horns on the ends. When it is cut open, one small, embryonic skate is revealed. In the sea, these cases will hold the embryos safely for several months and then split open to release the fully formed baby skates. (Rays give birth to live young.)

Sea Beans

The many different kinds of seeds that wash up on the beaches are called sea beans. After weeks or months at sea, many of these objects take on a very beautiful, smooth-polished appearance, and may be a real challenge to identify. Sea beans in Florida often come from the Caribbean, South America, and even from as far away as Africa, thanks to strong ocean currents.

Australian Pine

Sea Pearl

Lucky Bean

Mangrove

Hamburger Bean

Air Breathers, Water Breathers

Most mollusks obtain their oxygen from the water and cannot breath air. Even so, most can survive for several hours outside of the water by obtaining oxygen from small amounts of water which they retain within their shells and from the moisture available in the sand or other surroundings. This is how shells survive being stranded on sand bars and tidal flats at low tide.

Tropical Shells in North Florida

Apalachicola Bay in Northwest Florida is an unusual place which is very popular with shellers. Strong loop currents in the Gulf sometimes deposit the larvae of Caribbean shells in this bay. These shells then grow to maturity in an area which is well north of their usual range.

Coquina Rock

A number of places in Florida have outcroppings of "Coquina rock." This is a compressed mass of sand and shells (not exclusively Coquinas). It was used extensively for construction by pioneer Floridians. There are many old coquina rock structures in St. Augustine which are still standing today.

Sea Urchin

The spines of the Variegated Sea Urchin are short and are not dangerous to swimmers although some of the South Florida species have frightening spines resembling icepicks. It is not hard to imagine why sea urchins have been called the porcupines of the sea.

Most sea urchins are edible. They are especially prized by the Japanese who serve the dish at almost every sushi bar and call it "uni" (pronounced "oo-nee"). This food is also popular in the Florida Keys and the Caribbean islands. Only the gonads are eaten.

Tests of Variegated Sea Urchins

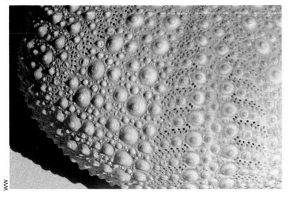

Holes and Bumps

Note the pattern of small holes in the shell through which the many tube feet of the live sea urchin protrude. The bumps on the shell are the places of attachment of the spines. This ball and socket arrangement allows movement of the spines which are pulled to and fro by muscles of the skin covering the shell. Dried urchins are sold in shell shops and are called "sultan's hats."

Bottom side showing mouth

△ The shell of a sea urchin is called a test and consists of ten wedge-shaped plates which are fused together. The mouth is a five-pointed apparatus very similar to that of the sand dollar.

The Variegated Sea Urchin camouflages itself by holding rocks and shells above itself with its tube feet. Thus it avoids the notice of crabs, fish, turtles and other predators. The rocks and shells may also act as ballast to help the urchin resist tides and strong currents. If you pick one up, it will immediately start to drop whatever it is holding.

◁ A sea urchin is stranded in a small pool of water remaining in the bay at low tide.

Mole Crab

Although it is often called a Sand Flea, this creature is really an unusual type of crab. Its movements, such as walking, swimming, and digging, are all backwards. The mole crab tumbles up the beach with an incoming wave. It very quickly digs itself into the sand (backwards, of course) and raises its feeding antennae upward into the outgoing wave to trap and strain plankton.

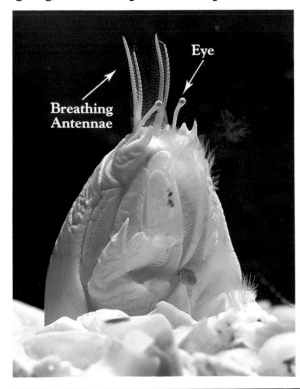

Mole Crab with feeding antennae extended into water

Mole crabs are used as bait by fishermen and are devoured by both fish and shore birds.

The mole crabs commonly seen on the beach are females. The males are much smaller and can sometimes be found clinging inconspicuously to the bodies of the females.

Red Tide

Red tide is a microscopic organism which sometimes appears in Florida waters during the warm months. When conditions are just right, this organism may "bloom" or multiply in such quantities that it kills large numbers of fish. It suffocates fish by depleting the oxygen in the water and also poisons them with toxins.

Red tide is almost a misnomer because the red coloration of the water is slight and rarely seen. However, the results of red tide are very obvious. Large numbers of dead fish appear on the beaches. Also, a person walking on the beach will experience irritation in the nose and throat, not only from the rotting fish, but also from gasses and other airborne chemicals released when the red tide organisms die and break up. Fortunately, red tide outbreaks are infrequent and usually only last a few weeks.

Oysters, clams and other mollusks which are filter feeders living in shallow water may be contaminated by red tide. They filter the red tide organisms from the water and thus become a concentrated source of the red tide toxin. Creatures such as shrimp and crabs which are not filter feeders are not much affected. During outbreaks of red tide, local health officials sometimes ban the harvesting of shellfish for a short period.

Sea Hare

Sea hares resemble the slugs found in gardens and are often called sea slugs. They are marine snails whose shells are small and internal rather than external. In the spring, many move into shallow water to deposit their tangled masses of egg strings.

Sea hares are often seen around jetties, seawalls, and rocks where they graze on algae. They propel themselves through the water with "wings" which are actually extensions of their foot. They appear graceful and beautiful in the water as their "wings" undulate rhythmically.

Why this animal was given the name "hare" is a mystery. While swimming in the water, two fleshy tentacles that vaguely resemble a rabbit's ears protrude from its head, but that is the extent of its physical similarity to a rabbit. It may be that the sea hare grazing on algae seems like a rabbit munching grass.

The Water Table in the Sand

There is usually moisture in the beach sand above the water level. This sand stays damp because of capillary action and because the water clings to the sand particles. Water vapor is also held between the grains of sand. This is very important to many beach creatures that burrow in the sand and might otherwise dry out at low tide. Also, wet sand has the property of becoming almost liquid if stirred up. This allows burrowing creatures to move through the sand much more easily.

Many of these sluggish creatures wash onto beaches and are helpless when stranded. If disturbed, they emit a harmless purple ink, like an octopus, which aids their escape. If you touch one with your foot for a few seconds, the ink will ooze out onto the sand. For this reason, the sea hare is also called the Ink Fish.

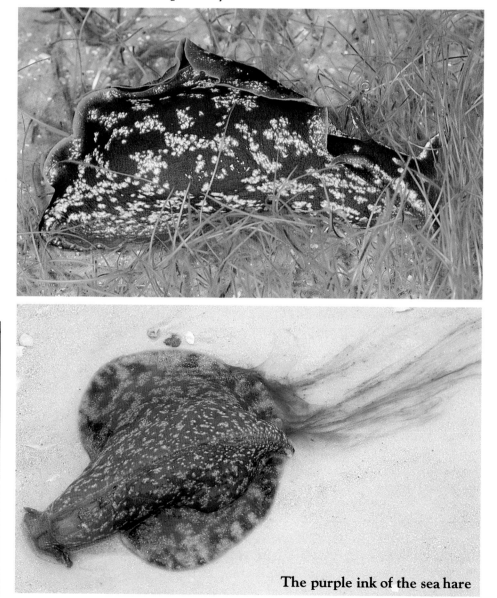

The purple ink of the sea hare

Starfish

Starfish are not at all related to fish, so scientists prefer the name sea star. A starfish has no head or tail nor right or left side. Its body is designed around five points, as are all echinoderms or "spiny-skinned" marine creatures. Most starfish have five arms or multiples of five.

On the underside of each arm are many tube feet with suction disks at the tips. Starfish feed on many of the two-shelled mollusks such as oysters and scallops and thus are a real threat to the shellfish industry. Starfish use their suckers to get a firm grip on the mollusks and apply pressure until the shells of their victims open slightly. Then the stomach of the starfish comes out through its mouth and enters the shell of the victim where it digests the prey "on location," so to speak, before retracting back into the mouth.

Live starfish among the grasses and rocks in Tampa Bay

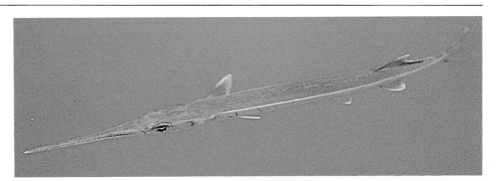

△ It is fascinating to place a starfish into an aquarium and watch the coordinated movements of its many tube feet as it climbs up the glass wall of the tank. The starfish controls the suction of its tube feet suckers by varying the water pressure inside the hollow tubes.

△ Beachcombers hoping to collect a perfect specimen of the Thin Starfish, or Nine-pointed Star, will be disappointed. As soon as this starfish is touched, it will drop an arm in the hope that the predator harassing it will settle for an easy meal and allow it to escape. After escaping, it will grow a new leg. The ability of starfish to regenerate lost limbs tantalizes scientists studying ways to repair the human body.

Needle Fish

Needle fish are commonly seen swimming on the surface of shallow water very close to shore and around docks. Children frequently amuse themselves by chasing these fish through the shallows. They are very fast and most of the time can easily escape a human's grasp, but if they are cornered, they will make some magnificent jumps and dives. Needle fish do not have enough meat to be worth eating but are used as bait for snook.

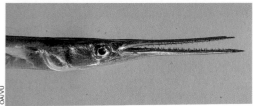

Needle fish are easy to see because they usually swim right on the surface of the water and they are a lot of fun to watch, especially at night around lighted piers.

Crabs

Ghost Crab

Ghost crabs are found on Florida beaches most of the year although they may remain dormant in their burrows for a few months during the cold season. Ghost crabs are not easy to dig up as their burrows may be 3 to 4 feet deep. On some Caribbean islands, ghost crabs are eaten with stuffing.

Ghost crabs become active when the sun is low or at night and are often seen scurrying to the surf to wet their gills. Although land-based, they cannot breathe air directly and must have access to sea water to survive. A special compartment (gill chamber) is kept moist and oxygen from the air is passed through the sea water on the gills.

The diet of the ghost crab consists of mole crabs and coquinas as well as the scavenged remains of dead fish and other creatures found on the beach, such as sea turtle hatchlings.

The mature ghost crab is rather yellow in color, but young ghost crabs are speckled white and grey and blend very well with the sand. When not running, the ghost crab will crouch down close to the sand to reduce its profile and its shadow. Sometimes it is barely visible, almost like a ghostly apparition, hence its name.

△ Ghost crabs are frequently seen wetting their gills in the surf at sunset time or at night.

△ Juvenile ghost crabs lack the yellow coloring of adults and blend with the sand.

The footprints of a ghost crab digging a burrow form a pattern in the sand.

Flame Crab

▷ Also called the Box Crab, the Flame Crab is a squarish creature that pulls its big front claws tight against its body. Parts of dead Flame Crabs are often found on the beach. Alive, it burrows into the sand with only its eyes and the top part of its claws protruding. The name Flame Crab comes from the flame-shaped, purplish-brown markings on the top of its shell.

Calico Crab

◁ This is one of the most beautiful crabs in Florida. Usually, only fragments of its shell are found on the beach. The Calico Crab is very active because its gills are unusually large and take in large amounts of oxygen.

Many Crab Species

Scientists have identified at least 70 different species of crabs along Florida's coastline. The actual number might be twice as large considering the many species that are small and easily overlooked.

No-See-Ums (Sand Gnats)

These tiny biting insects are common at many beaches, especially at night, or in daytime when the sun is low and the breeze has stopped. Technically, they are called midges. They are not impossible to see, just very hard to see because they are so small. To the naked eye, they appear as a tiny black dot, smaller than the head of a pin (and small enough to pass easily through screens). The intensity of their bites is astounding. Insect repellants help, but some people claim the best protection is a layer of baby oil (which may drown the little buggers).

The Best Shelling Spots in the World

The Philippine Islands are probably the best place in the world for shelling because of the warm water and many types of habitat. Inexpensive labor also makes this country the world leader in the commercial harvesting of collector shells and commercial shell craft production. The reefs along the northern and eastern coasts of Australia are also famous for their spectacular shells. The Caribbean islands, both coasts of Africa and many places in the Indian Ocean all have their own local specialties.

In Florida, Sanibel Island near Fort Myers is very highly regarded by shellers. The Florida Keys are also outstanding because these waters contain many Caribbean shells not found farther north.

Artist's vision of a No-See-Um

Sea Horse

Regarded as one of the most adorable sea creatures, the sea horse is a favorite with everyone. Its head is indeed very horse-like. It propels itself with a small fin on its back and even smaller fins on its sides, but must protect itself from currents by grasping underwater plants with its tail as it is not a strong swimmer.

Kangaroos of the Sea

One of the most remarkable facts about the sea horse is that the female deposits her eggs in a special pouch on the male's body. The male incubates the eggs, and eventually hundreds of tiny baby "sea ponies" emerge. Undoubtedly, many human females would like to be able to do this. (Note that for *real* kangaroos, it is the female that carries the babies in a pouch.)

Sea horses usually breed in quiet bays where they cling to sea grasses or other objects under water with their tails. They feed on very small marine life. If they lose their grip on a stationary object, currents can take them out into the open ocean. They are not strong swimmers, so storms and strong tides often wash them onto beaches where they are found in piles of sea weed at the high tide line.

Turtles

Many Florida beaches are visited at night during the summer months by endangered Loggerhead Turtles that lumber ashore to lay their eggs in the sand. These huge creatures (that can weigh over three hundred pounds) face many obstacles in reproducing their kind. Lights of condominiums confuse sea turtles. When their egg-laying is complete, the Loggerheads try to return to the sea. In the past, the direction of the sea was indicated by the glow of moonlight on the water. Now, the lights of condos beckon in the opposite direction, and confused turtles sometimes end up on roadways.

Civilization has brought a few small benefits. The greatest threat to turtle eggs used to be raccoons. Now the raccoons are too busy with scraps from the condo garbage cans to bother with digging up eggs. In spite of all conservation efforts, the fate of the Loggerhead is still uncertain.

Turtle hatchlings head for the sea

Parchment Tube Worm

These tubes are found on the beach in great numbers after storms. While the animal is alive, the tube is buried in a "U" shape with each end sticking up above the sand. The worm remains in its tube and sets up a flow of water, in one end and out the other. Thus the worm can filter particles of food from the sea water.

△ A beach littered with the tubes of Parchment Tube Worms.

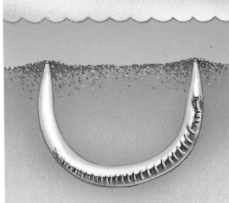

△ The tubes are buried in the sand with the ends open to filter water.

△ The much smaller casings of another species of tube worm, the Sand Grain Tube Worm.

Sea Squirt

The sea squirt is a sack-like creature that attaches itself to pilings or underwater plants. It has two siphon tubes so that it can take in sea water and strain out food. If disturbed it will squirt out its water, hence the name sea squirt. Some sea squirts live together in colonies and have a very different appearance (see page 66).

Crucifixes

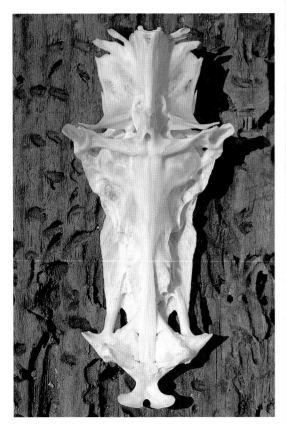

Art in the Sand

The community of Treasure Island near St. Petersburg sponsors an annual sand castle and sand sculpture contest which attracts participants from many states. Sand art is usually created by packing wet sand tightly and then carving it using the hands and small tools. One year at the annual festival, participants created a gigantic sand castle several stories high.

The Crucifix is not a shell, but a fish bone (part of the skull of a gafftopsail catfish or sail cat). It is found on beaches after the natural death of the fish or after fishermen have discarded it as an unwanted catch.

The Crucifix is said to resemble Christ on the Cross and seems to display the wound caused by the spear. If shaken, there may be a rattling inside which is said to be the sound of the dice thrown by the soldiers gambling for Christ's garments. This sound is caused by two small bone structures which help the live catfish keep its equilibrium in the water. It is said to be good luck to find a Crucifix that will rattle. The Crucifix may be cleaned by soaking it in bleach and then rinsing it thoroughly.

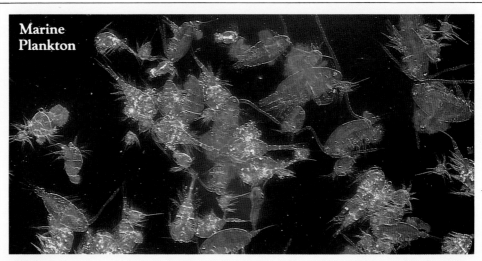

Marine Plankton

Plankton

Plankton are microscopic organisms which live in sea water. Some are plants and some are animals. They are at the bottom of the food chain and either directly or indirectly support much of the life in the sea.

Finding Shells on the Beach

Shells and seaweed are deposited along the high-water mark on sandy beaches in full view of the beachcomber. But there is another good place to look for shells: a small drop-off or lip of sand a few inches deep, right where the surf is breaking. Many shells never make it over this lip and can be collected by wading or snorkeling along the surf line and sifting through the quantities of shells that are continuously tumbled there by the waves.

Sponges

Sponges are not plants; they are animals that grow attached to stones, coral or other fixed objects on the bottom. There are many different species in Florida of various sizes, shapes and colors. The sponges that wash up on the beaches are simply the skeletons of formerly live animals. Beware of red or orange sponges as they may contain skin irritants.

△ Dead Man's Fingers is a sponge which is frequently found on Florida beaches.

Sulphur Sponge

Sea Pork
(not a sponge – see page 66)

The Greek Sponge Divers

At one time the little town of Tarpon Springs on Florida's West Coast had a thriving sponge fishing industry established by a group of Greek immigrants. Many boats went out daily carrying divers who wore weighted boots and heavy metal helmets with air lines. They harvested sponges while walking on the floor of the Gulf. After several natural disasters which damaged the sponge beds (plus the introduction of artificial sponges), this industry has diminished considerably but still continues today. Visitors can still buy natural sponges and take tours on the sponge boats.

1. Greek sponge diver with heavy metal helmet. 2. Diver surfaces amid mass of bubbles. 3. Suit is inflated with air and diver wears weighted shoes to walk on bottom. 4. Live sponge direct from the bottom of the Gulf.

Sea Whips

Also known as soft corals, these plant-like animals are usually dead when washed onto the beach, but occasionally they are found alive. If a live specimen is placed in a salt water aquarium, dozens of small polyps (each an individual animal) emerge along the stems of this colonial creature. Each little polyp animal gathers plankton from ocean water. The most common species washed ashore in Florida is called the Purple Sea Plume, although the color can vary from red or yellow to purple. Most sea whips anchor themselves to a shell or rock on the bottom. A number of live shells including Flamingo Tongues (page 73) attach themselves to sea whips and some may still be attached when the sea whip washes onto the beach. After a sea whip has been stranded out of the water for a while, its remains look like a thin piece of wire, hence its name.

◁ Note older, wire-like sea whip at top of photo.

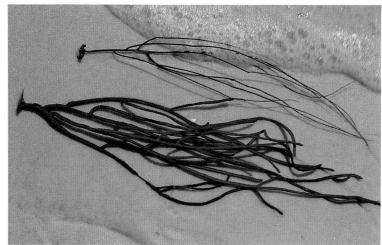

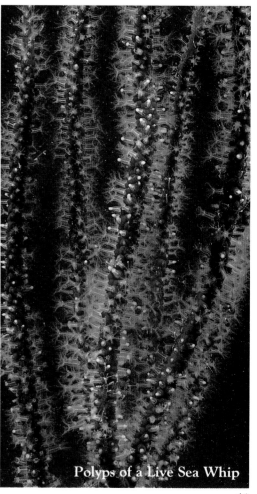
Polyps of a Live Sea Whip

Seaweed

Red Algae

Most plants which are called seaweed are either algae or marine grasses. Scientifically, seaweed is not a very accurate name because algae and marine grasses often grow in places other than the sea, such as lakes and rivers.

△ Algae appear on the beach in many shapes and colors.

Sargassum Weed

One of the most interesting kinds of seaweed in Florida is Sargassum Weed. It is carried by ocean currents primarily from the Sargasso Sea, south of Bermuda, to Florida, where it frequently washes up onto the beach. A clump of Sargassum Weed often contains a whole group of marine animals, all of which are brown to yellow in color, like the "weed" itself, and are perfectly camouflaged for a safe life within the tangles of this aquatic plant. Animals living in this colony may include crabs, shrimp, nudibranchs and the tiny Sargassum Fish.

Sargassum Fish

Sargassum Crab

△ Note the little "berries" on the Sargassum Weed. They are actually air bladders which allow the weed to float and travel great distances.

Vegetation in the Shallow Bays

Florida's grass flats are breeding grounds for small fish and are very important habitats for mollusks as well. Turtle Grass is the most important of the marine grasses, but there are also several others including Cuban Shoal Weed, Widgeon Grass and Manatee Grass.

Eel Grass is a catch-all phrase referring to all the different kinds of grasses found on the grass flats, but it is most commonly used in reference to Turtle Grass.

△ Turtle Grass growing in the shallows of Tampa Bay.

△ Piles of Turtle Grass on the beach.

What are those little white spots?

The Turtle Grass seaweed that is so common on beaches is frequently covered with little white spots. Under magnification these spots appear to be tiny spiral shells. They are actually hard-tube marine worms known as *Spirorbis spirillum* which attach to the grass and spend their lives filtering food from the water. Although they are not mollusks, they secrete an outer covering which looks very much like a miniature seashell and is made of similar material. When the seaweed is in water, the patient observer can see, extended from the spiral tubes, tiny tentacles similar to those of the much larger, but closely related, Feather Duster Worm (*page 91*).

△ The little white spots are marine worms.
▽ A hand lens shows the beauty of the *Spirorbis* spirals.

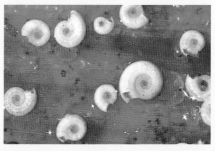

Sand Dunes

On certain beaches, dunes are nature's best defense against erosion by wind and sea. Dunes begin to form when grasses and seaweed are piled up by the surf at the high-tide line. When the wind blows, these piles trap sand. When enough sand accumulates, sea oats and other vegetation can take root and stabilize the shifting sands somewhat, forming what is called a fore-dune. The seaweed then serves as fertilizer for these soil-holding plants. Public officials who wish to have the beaches swept clean of seaweed must take into account the important role played by this seemingly useless debris.

Wind causes the dunes to gradually move up the beach as newer dunes are formed again at the high-tide line.

The result is a series of mounds which break the force of wind and storm tides and give a barrier island a fighting chance for survival.

△New dune is forming at bottom of photo.

Coral

The hard, limey part of coral is produced by very delicate creatures (polyps). They may be solitary or live in colonies. The stone-hard parts of coral are all that most people ever see of these small but beautiful organisms.

Broken Shell Identification Quiz

Can you name the shells these fragments represent? After dead shells have been tumbled in the surf, few arrive on the beach in good condition. Sometimes it takes real skill to identify the bits and pieces. Sharpen your knowledge with this quiz. The answers are at the bottom of the page.

Answers to the Broken Shell Quiz: 1. Slipper Shell 2. Sea Urchin 3. Moon Snail 4. Banded Tulip 5. Lightning Whelk 6. Fig Shell 7. Florida Fighting Conch 8. Cross-barred Venus Clam 9. Pen Shell 10. Calico Scallop 11. Quahog 12. Olive Shell 13. Surf Clam 14. Prickly Cockle 15. Van Hyning's Cockle

ROCKY BEACHES

Point of Rocks,
Sarasota, Florida

Top Shells

Top shells are so named because they resemble toy spinning tops. The live animals feed on seaweed. The shells have a pearly lining inside. The outer layer of some top shells can be removed with acid to reveal iridescent color underneath. Top shells and turbans that have been processed in this manner are often sold in shell shops. *(See also page 101 for the Superb Gaza, an unusually beautiful, deep-water top shell.)*

Jujube Top

Chocolate-lined Top

Adele's Top

Beautiful Atlantic Top

Live top shell

The Jujube Top is the most common top shell in Florida and the Chocolate-lined Top is the rarest among those in the photo. There are also some other deep-water Florida top shells (not shown here) that are very rare.

Chocolate-lined Top Shell *Calliostoma javanicum* Gmelin, 1791 • Jujube Top Shell *Calliostoma jujubinum* Gmelin, 1791 • Beautiful Atlantic Top Shell *Calliostoma pulchrum* C. B. Adams, 1850 • Adele's Top Shell *Calliostoma adelae* Schwengel, 1951

Chestnut Turban

The genus name *Turbo* means "resembling a toy spinning top," and indeed, these shells do have this appearance. Turban shells have interesting opercula, the trapdoors which cover the shell openings. The patterns on these opercula are useful in identifying the various species of turban shells. *(See page 107 for photos of some interesting turban opercula.)*

Chestnut Turban *Turbo castanea* Gmelin, 1791

Rock Shells

Rock shells are closely related to the murex. The shells are very variable in both shape and color.

Some of the Florida Rock Shells have knobby protrusions and some are smooth. They are usually found around oyster beds where they feed on young oysters by boring into them. They also eat mussels, barnacles and clams.

The Florida Rock Shells shown clustered on the piling are laying eggs. It is unusual to find them in such numbers.

Florida Rock Shell *Thais haemastoma floridana* Conrad, 1837 • Deltoid Rock Shell *Thais deltoidea* Lamarck, 1822

Florida Rock Shells

Deltoid Rock Shell with small chitons.

Florida Rock Shells laying eggs

Jewel Boxes

Spiny Jewel Box

The Spiny Jewel Box is distinguished by its long spines. When this shell is found along the beach, the spines are often worn down and less prominent. Its color is usually chalky white. It lives on rocks and rubble.

By contrast, the Leafy Jewel Box is very colorful. It is covered with abundant leafy protrusions (called fronds) all over its outer surface. The Leafy Jewel Box attaches itself to rocks, pilings and seawalls in shallow water. It is interesting to search for the point of attachment on the shell. Unlike the Spiny Jewel Box, it is seldom found on beaches.

▷ Jewel boxes are so named because they are shaped like containers. The lower valve (shell half) of the Leafy Jewel Box is rather deep inside while the top valve is flat like a lid.

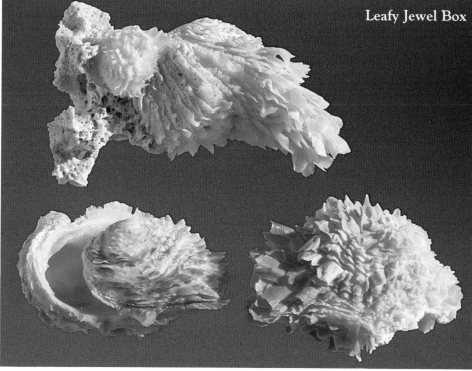

Leafy Jewel Box

Florida Spiny Jewel Box *Arcinella cornuta* Conrad, 1866 • Leafy Jewel Box *Chama macerophylla* Gmelin, 1791

Left from Right

The separate halves of the clam-like shells (bivalves) often curve in opposite directions, so they are not identical, but mirror images of each other. So, if you find two very similar shells that curve in opposite directions, it is likely that they are not different species or freak shells but just the left and right halves (or valves) of the same shell species. Scientists distinguish left from right by holding the shell so that the part of the shell above the hinge (called the beak, or umbo) is at the top and curving away from the viewer. For shells where the umbo does not curve, make sure the heart-shaped lunule is on the far side.

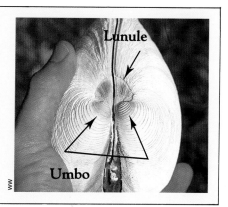

Lunule

Umbo

47

Worm Shells

Florida (white tip)

Two
West Indian
Worm shells
twisted
together

Fargo's (long spiral)

West Indian (short spiral)

△ These tubular casings once contained small marine snails, not worms, despite their worm-like shape. Worm shells are rather free-form. Their shapes are not as rigidly disciplined as most other shells. Notice, however, that the first coils are tightly wound and that later coils make much wider, looser turns. Another common name is Old Maid's Curl.

The worm shells in the top photo do not form colonies but live as lone individuals, unlike the other shells discussed below. They are not anchored in place and can move about freely.

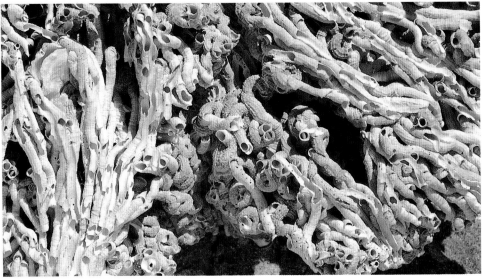

◁ The Irregular Worm Shells are heavier, darker, and more curled than the Variable Worm Shell. They sometimes attach themselves to colonies of Variable Worm Shells as well as to rocks.

△▽ Variable Worms live in colonies. Great masses of their shells joined together form a substance called "worm rock." The shells in the colony tend to line up parallel to one another.

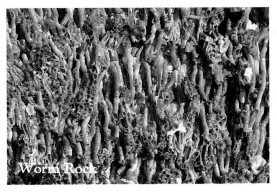

Worm Rock

West Indian Worm Shell *Vermicularia spirata* Philippi, 1836 • Irregular Worm Shell *Dendropoma irregularis* Orbigny, 1842 • Variable Worm Shell *Petaloconchus varians* Orbigny, 1841 • Florida Worm Shell *Vermicularia knorrii* Deshayes, 1843 • Fargo's Worm Shell *Vermicularia fargoi* Olsson, 1951

Sea Anemones

Undersea Flower Garden

Although resembling a flower, the anemone is actually an animal. It has a mouth surrounded by tentacles which can inflict a paralyzing sting to small sea creatures, but are usually harmless to humans. The anemone uses its tentacles to draw stunned creatures into its mouth where they are consumed.

Anemones usually attach themselves to something solid by means of a suction disk. If you pick one up, it may try to attach itself to your hand. Out of the water, the sea anemone will slump into a shapeless blob.

The common sea anemones found along the coasts of central Florida are mostly brownish to reddish in color, but in the Florida Keys anemones can be found in a variety of brilliant hues. The best place to observe anemones is in tidepools and in the small depressions of coastal rocks that stay filled with water when the tide recedes.

Watch This in an Aquarium

Many people have seen sea anemones in salt water aquariums. Most such aquariums will also contain the brightly colored clown fish* which can be seen

frolicking inside the anemone's "flower blossom." Clown fish allow themselves to be stung and even bite the stinging tentacles until they develop immunity. The clown fish is safe from predators when hiding among the stinging tentacles, and the Anemone eats the scraps from the clown fish's meals, a mutually beneficial relationship. Clown fish are only found in Pacific and Indian Ocean waters, but in Florida's coastal waters a similar role is played by the Anemone Shrimp (*see page 93*).

Clown fish, while not found in Florida waters, are bred for Florida aquarium shops and are frequently seen in local salt water aquariums.

Red Boring Sponge

The bright red patches that are very common on coastal rocks just below the high-tide mark are Red Boring Sponge colonies. This sponge serves a useful function in breaking down old shells into sediment, but it is very destructive to coastal rocks and breakwaters and has even damaged some Florida bridge pilings. It is able to penetrate such structures by secreting enzymes which dissolve the lime in the concrete.

DOCKS AND PILINGS

The Fouling Community

This unique marine group includes all the creatures that attach and/or bore. Entire books can be written about the relationships of the creatures that are found around docks and pilings. The community is divided into zones: above the high-tide line, below the low tide line, and the area in between. Each zone supports different types of animals. The fouling creatures do a lot of damage. Some can even bore into concrete or fiberglass. An entire industry has developed to defend marine structures against them by treating pilings with chemicals like creosote or painting boat bottoms with anti-fouling paint. On the positive side, these communities are ecologically important. A number of fish such as sheepshead, redfish and pinfish feed on the barnacles and oysters.

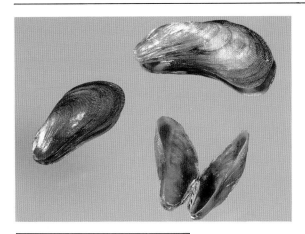

Mussels

Mussels are frequently found attached to pilings. The top-most mussels indicate the high tide line. They are also found on mud flats at the low tide mark and are associated with beds of *Spartina* (smooth cord grass). They can live with considerable pollution and actually prefer brackish waters.

Ribbed Mussel *Geukensia demissa* Dillwyn, 1817

Sea Roach

This insect (also called a Rock Louse) is a scavenger which very much resembles a household cockroach. Scientifically, it is called an Isopod. It avoids directs sunlight, and so it lurks in the shadows, and becomes more active around sunset or on cloudy days.

A Very Rude Habit (at least by human standards)

The Sea Roach can drink water in the usual manner through its mouth, but prefers to drink through its anus. While clinging to a piling, the Sea Roach lowers its rear end into the water, enlarges its rectum, and then takes in water by reversing the direction in which fluids normally flow through the body!

The Sea Roach can change color by controlling the concentration of pigments in special cells. It is usually much darker in color during the day.

50

Barnacles

Acorn Barnacles feeding

Barnacles are among the marine animals known as fouling organisms because they accumulate underneath boats and their removal is expensive. They maintain their hold on their support with a waterproof glue which is one of the strongest known to man. This glue is greatly envied by dentists and has been the subject of much research in efforts to extract it or duplicate it.

△ Watch the barnacle closely when it is under water and observe the feathery tentacles that whip out and back into the shell to strain food from the water. The part of the barnacle that attaches to its support is its head and the tentacles that kick out through the water to catch food are modified legs.

Rock Barnacles with small Lineolate Periwinkles

△ Barnacle shells have an interesting design. These shells are not made in one piece but are a series of six interlocking plates.

Baby barnacles drift through the sea searching for a place of attachment. Older, attached barnacles put out a chemical which attracts other barnacles. This is why barnacles are usually found in clusters. Clustering benefits barnacles by improving chances for fertilization of their eggs.

Shipworms and Boring Clams

So That's Why the Driftwood Is So Full of Holes!

Shipworms and boring clams make worm-like holes in wood but are not actually worms. They are tiny, two-shelled mollusks. Both are common in Florida. The marine industry has been at war with these creatures for hundreds of years because of the destruction they cause to ships and pilings. Some boring clams actually eat wood, others bore into it just for living quarters. On the positive side, they destroy floating lumber that could be a hazard to ships.

The boring clam digs its tunnels by opening and closing its shell in a chewing motion. Break open a piece of driftwood and you may find these little guys at work. Notice that the tunnels get larger in diameter as they go deeper into the wood. This is because the boring clams grow larger as the burrow progresses. The tunnels created by some species are lined with a hard coating. Some species of boring clams can even bore into coral rock.

Shipworms (not shown in photos) do much the same kind of damage as boring clams, but the creatures are

The working end of a boring clam.

Tunnels lined by boring clams.

Boring Clam *Martesia striata* Linné, 1758
Shipworm *Teredo navalis* Linné, 1758

different in appearance. They have long worm-like bodies which are many times larger than their small shells.

The Sun, the Moon, the Tides!

Tide charts show the times of high and low tides, and the size of the tides is indicated as the difference in feet, plus or minus, from the typical high or low tide. A chart may say, for instance, "Low tide: —0.7," meaning seven tenths of a foot lower than normal.

Why does the number of high and low tides vary from place to place, and even day to day in the same place? As the gravitational forces of the sun and moon move sea water on the surface of the earth, masses of land get in the way.

Water is forced through narrow passes and inlets. The distance water has to travel makes tides different at different locations. These factors create their own rhythms in the movement of the water, so the level of the high and low tides and even the number of high and low tides in a day is constantly changing. It is not unusual for one place to have four distinct tides on a certain day while a few hundred miles away there might be only two tides.

At full moon and new moon, the moon is in line with the earth and the sun. With the addition of the sun's gravity, the movement of water is increased and tides become stronger. Highs are higher and lows lower. These are called spring tides and allow some of the best shelling. The weakest tides (neap tides) occur at quarter moon and three quarters moon.

BAYS

Beachcombers: Take a Break!

Try wading on the mud flats of Florida's bays at low tide and experience a whole new world. Sandy beaches are littered with dead shells, but the ankle deep waters of the bays are literally crawling with live shells. Here you will easily find many species which rarely wash up on beaches. Just check your tide schedule, put on some old tennis shoes, and go. Causeways often provide the best access as well as convenient parking.

Periwinkles

Periwinkles are usually found out of water. They attach themselves to the roots of mangrove trees, pilings, and the leaves of Spartina grass where they feed on algae. The markings on the shell resemble the veins in the Spartina grass and act as camouflage. These snails climb up their chosen supports to avoid rising tides and descend at low tide to search for food on the mud below. Periwinkles in turn become food for crabs.

Some periwinkles live in the environment above the high tide line. Periwinkles are kept moist by spray but stay out of the water for most of their lives. They must, however, descend into the water when it is time to lay their eggs. *(see also the Lineolate Periwinkles on page 51).*

Angulate Periwinkle *Littorina angulifera* Lamarck, 1822

Coffee Melampus

These snails are abundant in the leaf litter of the red mangrove trees where the tides keep the soil moist. They retreat to the branches of the mangroves to escape incoming tides. Melampus snails have modified gills and can breathe in either air or water. Melampus snails are most active at night and during the sunrise and sunset hours.

Coffee Melampus *Melampus coffeus* Linné, 1758

Horse Conch

The Horse Conch is not actually a conch but a member of the tulip family. It is the official state shell of Florida, and obviously a really big fellow. It is probably called "Horse Conch" because it gets to be as "big as a horse." It is one of the largest shells in the world and can even damage the props of outboard motors. The Horse Conch will eat most any mollusk including its own species, but it especially likes to feed on pen shells. Collectors searching for the giant Horse Conch go to sandy grass flats where pen shells are numerous and can usually be sure of finding some Horse Conchs.

Live Horse Conch

The live animal is bright orange

Florida Horse Conch *Pleuroploca gigantea* Kiener, 1840

Color Changes

△ Horse Conchs are prized for their beautiful orange color, and the color of small Horse Conchs is the most intense. Some Horse Conchs (but not all) change to tan or cream color as they grow larger. Although many of the largest shells remain orange, these shells are still noticeably paler and less intense in color than the baby shells. Regardless of color, mature Horse Conchs are very popular with collectors simply because of their awesome dimensions. The world record Horse Conch is over 26″ in length!

The Price of Shells

Shelling in Florida is mostly for fun. The shells found on the beach are not likely to make you rich. After a storm, look for a Junonia worth perhaps $5.00. In the bays it is possible to do a little better with live shells. Large Angel Wings and Lightning Whelks retail for $10.00. Golden Olives may bring $20.00 to $30.00 and a Florida Horse Conch over 20 inches may be worth over $100.00.

For divers in the Florida Keys there is the possibility of an Angular Triton ($10.00) or a Lion's Paw ($50.00). Florida shell dealers must rely on volume sales of inexpensive shells. In other parts of the world, there are several species of cones and cowries that sell for thousands of dollars. Fulton's Cowrie sells for over $10,000.00.

In general, size, color, and especially quality are important in determining value. But in the end, any shell is only worth what you can get for it.

Mud Snails

These half-inch snails can be seen prowling over the sand and mud near the low-tide line of quiet bay waters. They can be found in large masses when feeding on dead fish or crabs.

Mud Snails have a very keen sense of smell which immediately alerts the animals to the direction of a food source. They respond to a dead fish in their vicinity the same way we would respond to the smoke from Joe's Barbeque.

The dead shells of Mud Snails are frequently inhabited by tiny hermit crabs.

Common Eastern Nassa *Nassarius vibex* Say, 1822

Mud Snails feeding on dead fish

Angel Wings

Angel Wings are favorites among shell collectors. They are closely related to the boring clams which destroy docks and pilings, but Angel Wings only bore into soft mud. In fact, they are the only type of boring clam that does not bore into something hard. There are several different species in the family and most are commonly called Piddock Clams.

Angel Wings construct their deep burrows by rocking back and forth to dig and squirting jets of water to flush away the debris, just like a modern mechanical drilling rig.

The Angel Wing is so delicate that it cannot be simply wrenched out of its burrow in the mud. Once a shell is located, the collector must carefully dig around and under the shell and lift it very gently with both hands to avoid breaking it. The delicacy of this very popular shell adds much to its beauty.

Collectors look for a little piece of shell called a bridge or hinge which, along with muscles, holds the two halves of the Angel Wing together. This is usually displayed with the shell. When Angel Wings die, their fleshy parts decay, so the dead shells washed up on the beach are never found in pairs because they lack the hinge teeth and ligaments which keep most other two-shelled mollusks intact even after the live animals are gone.

Next to the bridge is a small patch of material which shellers call the scab. Some collectors display both the scab and the bridge along with their shell.

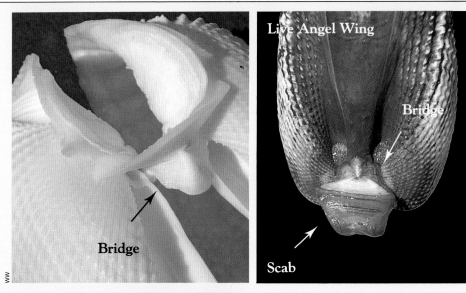

Bridge

Live Angel Wing

Bridge

Scab

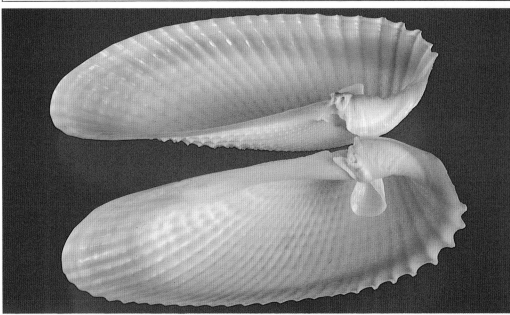

◁ Most Angel Wings are pure white, but the rarer (and more expensive) specimens have a touch of very delicate pink color over the interior. This special color is caused by local conditions in the shells' habitat.

Two other smaller species of Angel Wings are found in Florida — the Fallen Angel Wing and the Campeche Angel Wing. Neither is as prized as the beautiful species pictured here. Also in Florida we find the False Angel Wing, which is more closely related to the tellins but burrows into mud.

Live Angel Wings often inspire vulgar jokes among novice shellers because of the suggestive shape of their siphon tubes which can extend out from the shell 12 inches or more. Note that the live shell is partly covered with a black periostracum.

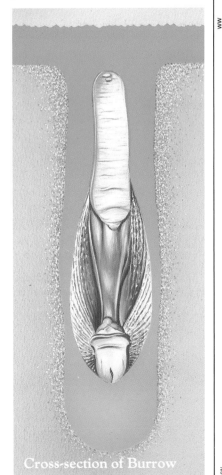

Cross-section of Burrow

Habitat: Shallow Bay Waters

Opening to Angel Wing Burrow

△ Shellers find Angel Wings by wading on the mud flats of the bays at low tide and looking for holes in the sand which indicate the presence of a burrow. Sometimes the siphon tube of the animal can be seen in the opening, but usually it is an inch or so below the surface and not visible.

If startled, the animal will retreat to safety at the bottom of its burrow, which may be as much as three feet deep. The poor creature is so fragile that it may even break its own shell during a hasty retreat. Walking on top of a burrow in the soft mud might crush an Angel Wing.

Spoon

◁ The small protruding part inside each shell is called a spoon. This projection is the point at which the muscles of the Angel Wing's foot attach to the shell.

Angel Wing *Cyrtopleura costata* Linné, 1758

57

Tulips

True Tulips
(three color forms)

Banded Tulips

Tulips are found almost exclusively in bay waters and are among the most popular collector shells.

Tulips are very carnivorous, and they are among the most aggressive and cannibalistic of all Florida shells.

▽ A True Tulip eating a smaller Banded Tulip. It is holding the smaller shell with its foot which also houses its mouth and stomach.

◁ The Banded Tulip generally is lighter in color and smaller than the True Tulip.

◁ If picked up, a live tulip, like a Fighting Conch, will extend its foot and thrash about violently, trying to right itself. It may even startle a novice sheller enough to cause him to drop the shell.

True Tulip *Fasciolaria tulipa* Linné, 1758 • Banded Tulip *Fasciolaria lilium hunteria* G. Perry, 1811

Crown Conchs

The Crown Conch, or King's Crown, usually has one or more rows of spines which give it a crown-like appearance although some specimens are spineless. It is usually found on mud flats near mangrove trees and oyster beds, but rarely on beaches. Individual shells may be very different in color, in length of spines, and the number of rows of spines. Notice that some Crown Conchs have rows of spines around both ends of their shells. Even rarer are pure white albinos.

◁ The two Crown Conchs at left show strong bands of color around their shell openings which indicate that they are good specimens. The shell on the right lacks this color, probably because of a different diet.

The live animal is black with white spots.

△ These Crown Conchs are almost invisible. Their camouflage covering of slime keeps them safe as they wait for the incoming tide. Most live shells can survive for several hours out of water because they breathe from a supply of water carried within their shells.

Crown Conch *Melongena corona* Gmelin, 1791

Pear Whelk

(not generally found on beaches)

The Pear Whelk is noted for its smooth surface and lack of the sharp points that are common to other whelks. It is smooth-curving from one end to the other and lacks the sharp angled "shoulder" which marks the widest point of most other marine snails. The Pear Whelk is often confused with the fig shell but lacks the cross-hatched texture of the fig shell. Left-handed specimens have been found but are very rare.

Pear Whelk *Busycon spiratum* Lamarck, 1816

Pear Whelk

Oysters

Oyster fishermen are much like aquatic farmers. They maintain oyster beds and provide support for the young, free-swimming oysters by dumping tiny, broken shells over the oyster beds. Slipper shells are often used for this purpose. These oyster beds are constantly under attack by starfish and oyster drills and are also vulnerable to disease and pollution.

Florida's oysters do produce pearls but not of substantial size or quality.

Oysters clinging to mangrove roots.

Young oyster larvae attach themselves permanently to a support and the oysters remain in the same spot for the rest of their lives. Oysters live in clusters, and low tide reveals many "oyster bars" or shoals covered with oysters which are above the water when the tide is low. Under Florida law, oysters must be more than three inches in length for legal harvesting.

Oyster bar

▷ The Coon or Frons Oyster is the most colorful of the Florida oysters and is found attached to mangroves and pilings.

Frons Oysters

Oyster Drill

△ These little shells prey on oysters by drilling holes through their shells with their rasping tongues. They cause millions of dollars of damage to commercial oyster beds and, after disease and pollution, are considered the oysterman's number one enemy.

Eastern Oyster *Crassostrea virginica* Gmelin, 1791
• Thick-lipped Oyster Drill *Eupleura caudata* Say, 1822
• Frons Oyster *Lopha frons* Linné, 1758

Cones

Alphabet

Sozon's

Crown

Carrot

Florida

Florida

Mouse

Stearns'

Jasper

Cones are very popular with collectors because of their graceful shape and beautiful colors. There are over 400 species worldwide and, certain species of cones are among the most valuable shells in the world.

Most live cones possess poison tipped "spears" or "harpoons" used to stab and immobilize prey, such as other mollusks or marine worms. The harpoons can · also be used as defense against predators like the octopus.

These weapons are formed from teeth of the cone animals and are attached to poison sacs. The harpoons (which are smaller than the tip of a ball point pen) are used once and then discarded. The cones simply grow replacements.

American cone shells are not very dangerous, but a few of the Pacific species produce some of the most virulent toxins of all animals. A number of people have died from these poisons.

△ Two species of cone shells are common in Florida, the Alphabet Cone and the smaller Florida Cone. The Alphabet Cone is known for its scribble-like markings which extend around the shell in a very orderly pattern. It is also called the Chinese Alphabet Cone by those who think the markings look like Chinese characters.

Sozon's Cone is a deep water shell that was once considered rare, but it is now salvaged in abundance from scallop fishermen. Notice the pointed spire and the distinctive brown and white markings.

It's a Shell-Eat-Shell World

The hinged shells (bivalves) are much more numerous than the spiral shells (univalves). Many of the spiral shells eat the hinged shells. They capture them and extract the meat from their protective shells with a variety of techniques including drilling holes, inserting muscle relaxing chemicals, and prying and wedging the shells open.

Florida Cone *Conus floridanus* Gabb, 1868 • Alphabet Cone *Conus spurius atlanticus* Clench, 1942 • Sozon's Cone *Conus delessertii* Recluz, 1843 • Jasper Cone *Conus jaspideus* Gmelin, 1791 • Stearns' Cone *Conus jaspideus stearnsi* Conrad, 1869 • Mouse Cone *Conus mus* Hwass, 1792 • Crown Cone *Conus regius* Gmelin, 1791 • Carrot Cone *Conus daucus* Hwass, 1792

OTHER BAY LIFE

Crabs

Fiddler Crabs

Fiddler Crabs feed in great masses on mud flats at low tide. They mass for the same reason that fish school, safety in numbers. A predator is less likely to attack a large army of moving, rustling fiddlers than it is to strike at a lone individual.

Fiddlers live in holes in the sand close to the edge of the water. They are usually found in large colonies and resemble armies on the march when they emerge from their burrows. These burrows are about 10 to 15 inches deep with a small chamber at the bottom. During high tide, the fiddler plugs the opening to its burrow and waits inside for the tide to recede.

Why "Fiddler?"

Fiddler Crabs are so named because the large claw of the male is thought to resemble a fiddle. The male fiddler uses this single large claw to attract a mate, by waving it in a certain pattern, and for defense during mating season battles with other males. The big claw is too large to be useful in feeding. Female fiddlers have two small claws of equal size.

Many shore birds prey on Fiddler Crabs. The Yellow-crowned Night Heron will wait patiently beside the opening of a burrow for its occupant to emerge. The Ibis has no need for such patience. With its long, curved beak, it can reach right into the burrow and extract the fiddler.

Male

Female

△ Fiddler Crabs may seem to eat mud, but actually they strain particles of food left in the wet sand by the receding tide. The residue from each strained claw full of sand is a small pellet. Note that fiddlers make two types of mud pellets, one type from feeding and another from burrowing. Those created by feeding are smaller. The pellets created by burrowing are usually found in patterns around the burrow opening.

Color Rhythms and Color Riddles

Fiddlers often feed at night. But a fiddler observed with a flashlight in the dark might be quite different in color than it would be in the daytime. The fiddler becomes lighter in color at night and darker during the day. This phenomenon is well known, but what function this color change might have is the subject of current research. Various theories suggest that the dark coloration in daytime may help protect the crab from ultraviolet radiation from the sun, aid in regulating temperature, or add camouflage.

◁ Fiddler Crab Quiz:
Can you spot the female?

Crabs

Horseshoe Crab

The Horseshoe Crab is not actually a crab at all but is related to a very ancient kind of animal called a trilobite, and it really looks the part. The long, fearsome-looking tail is not a weapon but is used like a rudder to steer the crab through the mud. It is also used for burrowing and helps the crab flip over if it accidentally gets turned on its back. Horseshoe Crabs are used in medical research to study the binding of oxygen to blood cells. They possess large amounts of blood which is bluish in color and significantly different in chemistry from the blood of mammals.

Horseshoe Crabs are very common along the Gulf beaches in the spring when they come onto the shore to mate and lay eggs. Large numbers appear at the time of the highest tide of the full moon. The females lay their eggs in the sand just above the high tide line. Since the water will not reach this level again until the next full moon, the eggs incubate in the sand for 28 days. On the high tide of the next full moon, the larvae hatch out and enter the water.

During the mating season, the smaller male Horseshoe Crab attaches itself with small hooks to a female.

It is common to see the larger female crab moving about with one or more males firmly attached to her shell. The male remains with the female until she lays her eggs. The male fertilizes the eggs and then goes his own way.

Mangrove Crab

This crab is usually found in the mangrove trees, out of the water. When alarmed, it releases its hold on a branch and drops straight down into the water — quite a surprise if one suddenly lands on your head.

Female with males attached. Note that Horseshoe Crabs have four eyes.

Horseshoe Crab attempting to right itself

Horseshoe Crabs are scavengers on the ocean bottoms. They have no teeth and must use modified leg appendages to chew their food. They have four eyes, two large compound eyes on the top of the shell and two smaller simple eyes close together on the front of the shell. A third pair of eyes is present on the juveniles.

Mangrove Crab

Blue Crab

Stone Crab

Stone Crab

◁ Stone Crabs are drab and not particularly attractive but are famous for the delicious meat of their claws, the only edible part.

Blue Crab

◁ This is Florida's most common edible crab. When it is collected between moults and before its new shell has hardened, it is called a soft-shelled crab, but the meat inside is the same.

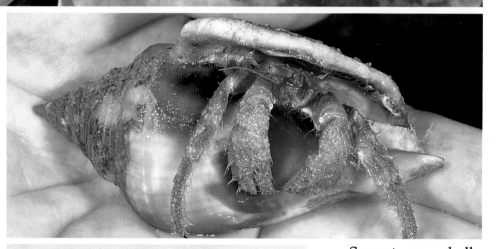

Large marine hermit crab in shell of whelk. The old whelk shell is covered with slipper shells.

Hermit Crabs

There are at least a dozen species of hermit crabs in Florida. Some are very large and some are so small they can hardly be seen with the naked eye. Most live in the sea, but in the Florida Keys there is a species of land-dwelling hermit crab which is an air-breather, although it still uses gills and must wet its gills frequently to keep them moist (*see page 90*).

Hermit crabs take up residence in the empty shells of various marine snails, holding themselves in their adopted shells with their hooked tails. When a hermit crab grows too big for its present shell it finds a larger one, leaves its old shell, and backs into its new home.

Why the Hermit Crab Needs to Borrow a Shell

Most other crabs get by very nicely with their own hard shells, but the body of the hermit crab is only partially covered with hard shell. Its abdomen is bare and unprotected. This is why the little hermit must "cover its tail" by finding an empty shell and backing into it. Hermit crabs have been known to live in man-made objects such as small bottles as well as shells!

Dreaming of the "House Beautiful"

Sometimes shells found along the beach contain hermit crabs. If your shell seems too heavy, place the shell in sea water to see if a hermit crab comes out. Marine hermit crabs cannot live for long out of sea water, so if you accidentally take home a shell containing a hermit crab, you will be notified of your mistake a few days later by the odor of the poor, deceased animal.

Sea Pork

(Tunicates)

These jelly-like blobs are common in bays and are sometimes found on the beach, especially after winter storms. In Florida there are many different shapes and colors. They are actually large colonies of hundreds of tiny animals bound together with a rubbery sheet of colored gelatin. They are not as slimy as they look and are harmless to touch. The colony attaches to some support and survives by straining plankton from the water. The tiny animals can be viewed by cutting open the colony and closely observing a thin slice against a strong light.

Sea Squirts

Sea Pork:
Amaroucium growing on pen shell

Golden Star Tunicate

Another name for these creatures is tunicate since their external covering is like a tunic. The scientific name is ascidian. Some types are called sea squirts because they will squirt water if picked up and squeezed (*see also page 39*). Most have siphon tubes for taking in and expelling sea water, and they use this pumping facility to filter out food.

Ambergris: The Pot of Gold at the End of the Rainbow

Every beachcomber's dream is to find a hunk of ambergris and become rich. Ambergris is a substance formed in the stomach of sperm whales. When sperm whales eat squid, the sharp, indigestible beaks of the squid, which could damage internal organs, accumulate in their stomachs. The whales' stomachs coat these accumulations with ambergris to allow them to pass through. The gray-colored blobs of ambergris float in the ocean and sometimes wash up on beaches. Amber and gris (the French word for gray) refer to colors. Fresh ambergris is gray, but it becomes amber-colored after floating in the water. Ambergris is used by perfume makers to preserve fragrances. A few years ago it was worth more than gold. It is now worth perhaps $150 per pound (but lumps found on the beach sometimes weigh 15 pounds). It has been found in the Bahamas, so why not Florida?

Winter storms in Florida dislodge the tunicate *Amaroucium* (shown at right) from coastal bottoms and hunks of the stuff wash onto beaches. According to a marine biologist with the Florida Bureau of Marine Research, every

A form
of Sea Pork
often mistaken
for ambergris

year several people come to his office to have chunks of *Amaroucium* identified in the hope that it will prove to be ambergris. One woman, who had apparently already mentally spent her new-found fortune, passed out upon hearing the true nature of the find.

Marine Worms

There are many kinds of marine worms and although they are very common, most tend to stay out of sight. In the bays, their presence is indicated by shapely piles of sand castings created as the worms burrow through the sand.

△ The casings of tube worms are frequently found on beaches, sometimes encrusted with small shell fragments. In bays, these tube casings are found sticking up out of the sand. If a casing is broken open, the worm will be found inside. Some species are used as bait.

▽ The open end of a tube worm extending above the sand to filter food from the water.

△ Burrows and sand castings of Acorn Worms.

Sand casting of marine worm

Sea Cucumber

Sea cucumbers are sausage-shaped animals which crawl through the sand of the grass flats. Some are inconspicuous in the mud because they are small, while others are over a foot long. One species in the Indian Ocean grows to over six feet in length. Sea cucumbers are sometimes washed up and stranded on beaches. These creatures strain small bits of food from mud.

The sea cucumber has an interesting defense mechanism. It can expel part of its guts when threatened. The attacking animal may swallow these parts or simply be confused by them, allowing the sea cucumber to escape and regenerate the internal organs that were sacrificed.

Sea cucumbers are echinoderms, related to sand dollars and sea urchins. They have the same radial symmetry as these other relatives, but to observe it, you have to look at the creature from the end (*see page 29*).

There is a tiny fish called the Pearl Fish which makes its home in the anus of some of the larger sea cucumbers. This little fish backs into the opening and waits in safety for its prey.

SPECIALTIES OF THE KEYS

www

Shelling in the Keys

For the ardent sheller, this is really exciting territory. But the terrain is rather different here. Sandy beaches are short, rather narrow, and alternate with coral rock shorelines. For the best shelling, it is necessary to wade in the water, walk the endless shallow flats at low tide, or even better, explore the clear, blue-green waters with snorkel or scuba gear. This is the "Caribbean" USA, and a whole new world of beautiful shell species awaits. All that is required is a little experience in spotting shells in their many natural habitats.

A Fortunate Location

Florida shells come from two different zones: the Caribbean Marine Faunal Province, and the Carolinian Province. Shells found in these two zones are very different, yet the two zones overlap in South Florida, giving Florida shellers the best of both worlds.

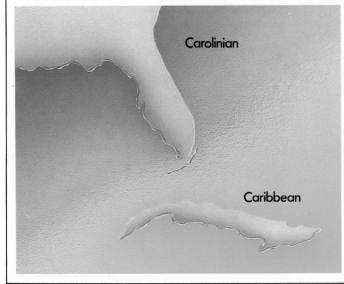

Trivia Shells

These tiny (1/2 inch) shells resemble cowries but are only distantly related and not members of the cowrie family. Note the three pairs of black spots which distinguish the Coffee Bean Trivia.

Four-spotted Trivia *Trivia quadripunctata* Gray, 1827 • White Globe Trivia *Trivia nix* Schilder, 1922 • Coffee Bean Trivia *Trivia pediculus* Linné, 1758

Coffee Bean   White Globe

Four-spotted

Angular Triton

This exciting shell has an unusual triangular shape when viewed from the end and is quite variable in color. It is a very tough shell due to its heavy ribbing. In the Keys, Angular Tritons are usually found in colonies. The live shells are covered with a hairy skin (periostracum) which flakes off after the animal has died and its shell has been cleaned.

Tritons are rather large shells which can reach 8 inches in length. They are usually found on shallow grass flats.

Angular Triton *Cymatium femorale* Linné, 1758

Three color forms of the Angular Triton

Vase Shells

The live shell is covered with a heavy, dark periostracum which may be hard to remove, although the shell itself is pure white. Most collectors leave the periostracum on the shell. This species is fairly common and is found on coral rubble and shallow grass flats where it eats worms and small bivalves. Where there is one, there are usually many.

Caribbean Vase Shell *Vasum muricatum* Born, 1778

Conchs

Pink Conch

Pink Conch

△ The foot of this shell (also called the Queen Conch) is the source of the famous Key West conch chowder, which was enjoyed from the time of the earliest settlers until recently, when this shell was placed on the endangered species list. Collection is now prohibited in the Florida Keys. Conch chowder is still served, but with conch meat imported from the Caribbean islands.

The Pink Conch is famous for the extraordinarily beautiful pink color inside the shell. There is a relation between the color in the shell and the type of food that the animal eats. Pigment is obtained from food and enters the blood of the shell animal. It is concentrated by the mantle (the part of the animal nearest the shell) and

Milk Conch

plays a role in the chemical process which forms the shell.

Pink Conch Strombus gigas Linné, 1758

Milk Conch

Note the very thick, heavy shell. The name, Milk Conch, is due to its overall white color, especially inside. But in the Keys it can also be orange, red, yellow, brown, or tan. It lives on coral rubble or grass flats and is most frequently found on the Gulf side of the Keys. Large groups of Milk Conchs have been seen by divers in deep water off Everglades City and Fort Myers. It is considered good to eat, like the Queen Conch, and is very abundant in the Caribbean where its shell is almost always white.

Milk Conch Strombus costatus Gmelin, 1791

▷ Notice that all true conchs have a notch on the lip of the shell. This is the place where the right eye-stalk of the animal extends out from the shell (*see page 19*). (see page 19) Scientifically it is called the stromboid notch.

The stromboid notches of various conchs.

70

Rooster-tail Conch

Rooster-tail Conch

This is a rare and highly-prized shell of the Keys. It is distinguished by the long extensions of its flaring lip. The few shells that have been found were probably collected at night. The Rooster-tail Conch is more common in the lower Caribbean islands.

Rooster-tail Conch *Strombus gallus* Linné, 1758

Hawk-wing Conch

The scientific name is *Strombus raninus*. "*Raninus*" in Latin means "little frog," a reference to the shell's warty back. It is found in shallow water throughout the Keys and comes in many beautiful colors.

Hawk-wing Conch *Strombus raninus* Gmelin, 1791

Hawk-wing Conchs

Fighting Conch (for comparison)

Pink Conch in surf

The Meaning of "Conch"

The word conch (pronounced "konk") is used to describe many large shells, but the only true conchs are those shells belonging to the genus *Strombus*. Oddly, the term conchology means the study of all shells (but not the live animals these shells once contained). There are many conchology clubs in Florida and the word is frequently seen on bumper stickers (such as "Suncoast Conchologists"). In the Caribbean, conch usually refers to the Queen Conch (Pink Conch) which is so popular for chowder. In Spanish, one word for shell is "concha."

The early settlers in the Keys were called "conchs" because of the popularity of the conch for food. Some present day residents of the Keys have declared themselves independent as the "Conch Republic" and have applied for admission to the United Nations.

Helmet Shells

Queen Helmet

Helmet Shells have unusually broad, triangular shaped lips adjoining their openings. Scientifically, these lips are called parietal shields.

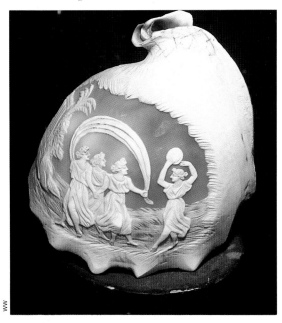

△ Helmets are famous as the shells from which cameos are carved. The helmet shell is formed in layers of different colors, and cutting through an outer layer will reveal a contrasting color underneath.

Queen Helmet *Cassis madagascariensis* Lamarck, 1822
• King Helmet *Cassis tuberosa* Linné, 1758

King Helmet

The Queen Helmet (also called the Emperor Helmet) is the largest of the three American helmets and can be over a foot in length. Helmets feed on sea urchins and sand dollars, a rather rough fare. The Queen Helmet lives in deep water and is usually found by dredging.

The King Helmet and the Queen Helmet look rather similar. The major difference is the shape of the wide lip, or shield. The King Helmet has a very triangular shield, while the shield of the Queen Helmet is more rounded. The Queen Helmet may grow much larger than the King Helmet.

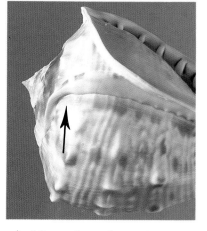

△ Note that the remnants of lips formed during earlier spiral growth are still visible.

Flamingo Tongue

This shell is famous for the beauty of the live animal which is covered with yellow-orange spots, each encircled by a dark ring. The Flamingo Tongue has been called the leopard of the sea. This animal feeds on sea fans and other soft corals at various depths.

Note the small beady eyes and long feelers on each side of the feeding snout. The live animal almost completely encloses its shell.

Fat Man's Belt
▽ The small Flamingo Tongue shells are not as spectacular as the live animal but are still quite attractive because of their unique shape. They feature a bulge around the middle with a raised ridge encircling it like the bulging belt of a very fat person.

Flamingo Tongue *Cyphoma gibbosum* Linné, 1758

Live Flamingo Tongue

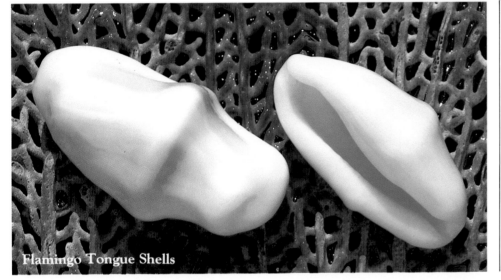
Flamingo Tongue Shells

Channeled Turban

△ This beautiful turban shell is found in the lower Keys near reefs and offshore rocks. Not really rare, it is nevertheless a valued collector's item.

Channeled Turban *Turbo canaliculatus* Hermann, 1781

Tiger Lucina

◁ This large bivalve is probably the most conspicuous shell along the shores of the Keys and on the shallow grass flats. It is often found as a hinged pair. The pure white exterior is decorated with a fine woven texture. The interior is ringed with purple and sometimes washes of yellow.

Tiger Lucina *Codakia orbicularis* Linné, 1758

Margin Shells

These shells are found in shallow waters in every type of habitat and are famous for their fabulous color, high gloss, and the beauty of the living animals. Margin shells are generally small, usually less than half an inch in length. To appreciate their beauty, the collector must look very closely and perhaps use a magnifier.

There are at least 14 species in Florida waters about half of which are found in the Keys and the others in bay waters farther north along both coasts.

Margin shells are so named because the adults have a thickened border surrounding the lip of their shells. They feed on scraps or fresh meat. They also attack other, sometimes much larger, live shells.

Orange Marginella

Live Margin Shell

Velie's Marginella shells feeding on Pen Shells

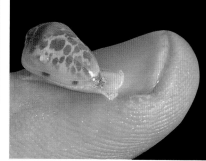

△ The Common Atlantic Marginella crawling on the finger is easily identified by the reddish spots around its "margin," two on each side, one at the front, and one at the back.

Orange Marginella *Marginella carnea* Storer, 1837 • Common Atlantic Marginella *Marginella apicina* Menke, 1828 • White-spotted Marginella *Marginella guttata* Dillwyn, 1817 • Velie's Marginella *Hyalina succinea* Conrad, 1846 • Orange-banded Marginella *Hyalina avena* Kiener, 1834 • Golden Line Marginella *Marginella aureocinta* Stearns, 1872 • Snowflake Marginella *Marginella lavalleeana* Orbigny, 1842 • Teardrop Marginella *Granulina ovuliformis* Orbigny, 1841

White-spotted

Common
Atlantic

Velie's

Orange

Orange-
banded

White-spotted Marginella

eyes

mantle
covering
shell

siphon

foot

tentacle

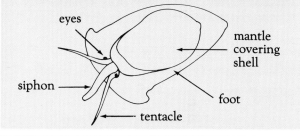

Teardrop

Snowflake

Golden
Line

△ Three marginellas on a finger tip.

Florida's Tiny Shells

Nature seems to favor small shells. Of the 1500 or so species of shells in Florida waters, perhaps more than 700 are smaller than the margin shells. Worldwide, more than half of all shell species are less than an inch in size. Most of these are much smaller and require magnification to observe and appreciate. Although the really tiny shells are popular with only a few collectors, they are of interest to researchers. For the marine scientist seeking immortality, this field offers the best opportunity to discover and name a new seashell species.

Star Shells

The Long-spined Star Shell is popular because of its unusual flattened shape and beautiful low spiral and spines. Live shells are found on shallow sandy, grassy areas.

By contrast, the American Star Shell is not flat at all, but has an unusually high spire.

Long-spined Star Shells

Live American Star Shells on rock

The scientific name of the Long-spined Star Shell, *phoebia*, comes from the name of the Greek god Phoebus Apollo or "radiant apollo" and refers to the spines of the shell which resemble a child's drawing of the sun's rays.

The Long-spined Star Shell ranges along the Florida West Coast as far north as the Panhandle. The shells in certain areas develop a beautiful green and gold coloration.

American Star Shells

Long-spined Star Shell *Astraea phoebia* Röding, 1798
• American Star Shell *Astraea americana* Gmelin, 1791

Netted Olive

The Netted Olive is not as common in the Keys as it is in the Bahamas. It is sometimes found along the Atlantic Coast. Many of the shells sold in shops have been dredged from deep water.

The Netted Olive is glossy like the Lettered Olive and is usually pale overall with tinges of pink and white.

Netted Olive *Oliva reticularis* Lamarck, 1810

Lettered Olives

Netted Olives

AROUND ROCKS IN THE KEYS

△ Snorkeling along the rocky shore in search of shells.

△ Looking under a big rock for a big cowrie.

False Cerith

This shell is extremely common along the shoreline and can be seen at low tide, clinging to the rocks in thick clusters. False Ceriths are not exclusive to the Keys but can be found northward as far as Sarasota, although in smaller numbers. The coloring of the shell varies but usually includes black or white spiral bands. That famous Caribbean bird, the flamingo, dines on these shells.

Caribbean False Cerith *Batillaria minima* Gmelin, 1791

△ Live False Ceriths are found in dense clusters on rocks.

Dove Shells

These small (less than 1/2 inch) shells are usually found under rocks and rubble in the daytime, but crawl about at night to scavenge for food on sandy or muddy areas. They are also found on seaweed. They are common and easy to find in all tropical waters. The family is complex with many kinds of shells and many colors.

Common Doveshell *Columbella mercatoria* Linné, 1758 • Well-ribbed Doveshell *Anachis lafresnayi* Fischer & Bernardi, 1856 • Rusty Doveshell *Columbella rusticoides* Heilprin, 1887 • Semiplicate Doveshell *Anachis semiplicata* Stearns, 1873

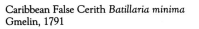

Limpets

The limpet is one of the few single-shelled marine snails (gastropods) whose shell is not coiled. Because of its shape, this shell is also called the Chinese Hat. The live animal has teeth which it uses to scrape algae from rocks. The hole at the top of the "hat" is the exit for water and waste material. The famous "keyhole" is not at the top in baby shells, but appears as a slit near the edge of the shell. As the shell matures, the opening moves until it is finally positioned at the very peak.

Some limpets grind away at rocks until they create a small depression into which their shell will fit very snugly, giving them a much better grip and protection against the waves. They roam about grazing on algae, especially at night, but return to the same cavity. Collectors must use a thin tool or knife blade to pry them from rocks. Because of their grinding, limpets take some blame for the erosion of rocks along the shore. The shape of limpet shells, like that of barnacles, is quite streamlined and ideal for deflecting the forces of countless pounding waves. The heavier the surf in a particular area, the flatter will be the shells of the limpets living there.

Cayenne Keyhole Limpet *Diodora cayenensis* Lamarck, 1822

Live Limpet

Should the strongest
　arms endeavor
The limpet from
　his rock to sever,
'Tis seen its loved
　support to clasp
With such tenacity
　of grasp,
We wonder that such
　strength should dwell
In such a small
　and simple shell!

William Wordsworth

The Color of Sea Water

Florida visitors often notice that the sea water in different parts of the state is different in color. Even the water in one location can appear very different from day to day.

Pure water is not really clear but is slightly blue in color. This blue color is not strong enough to be noticed unless you are looking through a lot of water. For this reason, water in a drinking glass seems clear, but the water in a swimming pool appears blue.

Green color in sea water is caused by plankton in the water or grasses on the bottom.

The color of the sky, the amount of solids in the water and the angle of the sun all have a strong bearing on the appearance of water in the sea.

The incredible blue and green colors of waters around the Florida Keys are due in part to the presence of lime in the water. Coral reefs are made of lime and the abundant presence of reefs in the Caribbean area makes these waters especially striking in color.

Cowries

Deer Cowrie

Measled Cowrie

Atlantic Gray Cowrie

Atlantic Yellow Cowrie

Cowries have symbolic meaning to people in many parts of the world. In addition to representing long life and good luck, the cowrie has sexual connotations because of its resemblance to the female reproductive organ. In Japan, women giving birth hold cowries in their hands in the hopes of easy delivery, and in the Japanese language the cowrie is called "koyasu gai," the "easy delivery shell." The ancient Egyptians used cowries for the eyes of some of their mummies to ensure good vision in the afterlife.

Cowries are easily found among the rocks of the shoreline near the highway in the Florida Keys, but they can also be found farther north by scuba divers. North of the keys, cowries inhabit rocky ledges offshore in about 30 feet of water, but they are not found among the rocks of the shoreline, and dead cowrie shells rarely wash up on beaches.

Cowries come out at night to feed on algae. They are often found by collectors searching the rocky shorelines of the Keys with lights.

Atlantic Yellow Cowrie *Cypraea spurca acicularis* Gmelin,1791 • Atlantic Gray Cowrie *Cypraea cinerea* Gmelin, 1791 • Deer Cowrie *Cypraea cervus* Linné, 1771 • Measled Cowrie *Cypraea zebra* Linné, 1758

The Deer Cowrie is not only the largest cowrie in Florida but the largest cowrie in the world. It can grow to a length of seven inches. Measled Cowries may resemble small Deer Cowries but the Measled Cowrie has spots along the edge of its base which have dark centers and resemble eyes.

The "Living Coat" of a Measled Cowrie.

Why Some Shells Are Very Glossy

Olives, moon snails, cowries and margin shells, to name a few, have very shiny, porcelain-like outer surfaces. The soft bodies (mantles) of these marine snails extend out from the openings of their shells and secrete glossy layers on the outside of the shells. The mantles of the live animals also partially cover the outside of the shells. This protects the shells from abrasion and the attachment of barnacles, sponges, and corals. The movement of the soft mantles over the exterior of these shells constantly cleans and polishes the glossy outer surfaces.

Bleeding Tooth Nerites

Nerites

The Bleeding Tooth is found in a variety of colors. It gets its name from the red stains around the conspicuous teeth seen in the shell opening. There are usually two teeth and one is usually larger than the other. This shell is popular with novice collectors who often take home a specimen to give to a dentist friend.

Bleeding Tooth Nerites are found on rocks in the splash zone of the shoreline. They have become difficult to find in the Keys because of over-collecting but are common in the Caribbean Islands.

Zebra Nerites

Four-toothed Nerites

△ This shell is often found feeding on algae in tidal pools or on grass flats.

◁ The rocky habitat of a Four-toothed Nerite.

Emerald Nerites

△ The Emerald Nerite is one of the world's few green shells. This tiny shell feeds on marine grasses. It is difficult to locate and collect because of its small size. In the Caribbean, it is often found in the beach debris at the drift line.

Emerald Nerites with grain of rice

△Emerald Nerites showing details of their fine markings with grain of rice for size comparison.

▷ Part of the charm of Virgin Nerites is that no two shells are alike. They are found on mangrove roots in brackish water, but they are more common on mud flats.

Bleeding Tooth *Nerita peloronta* Linné, 1758 • Zebra Nerite *Puperita pupa* Linné, 1767 • Emerald Nerite *Smaragdia viridis* Linné, 1758 • Virgin Nerite *Neritina virginea* Linné, 1758 • Four-toothed Nerite *Nerita versicolor* Gmelin, 1791

Virgin Nerites

Purple Sea Snails

The live Purple Sea Snail makes a raft of small bubbles and uses this raft to float on the surface of the ocean where it feeds on the Portuguese Man-of-War.

Purple Sea Snails are found world-wide in all tropical and subtropical waters. They live in large numbers in the open ocean and sometimes wash ashore in mass, especially in the springtime after a storm with strong winds. There have been times when the shores of Key West actually appeared purple because they were so littered with these shells.

Purple Sea Snail *Janthina janthina* Linné, 1758

Chiton

Chitons (pronounced "ky-tuns") are among the most primitive mollusks. Their shells are made of overlapping plates, like a suit of armor, hence another common name, Coat-of-Mail Shell. The plates are banded together at the edges by a girdle. Chitons usually cling to rocks or dead coral, although they are sometimes found attached to pen shells that have washed onto the beach. If you pry one loose, it will immediately curl up into a ball for defense.

The individual plates of chiton shells are sometimes found on the beach and can be recognized by their butterfly shape.

Fuzzy West Indian Chiton *Acanthopleura granulata* Gmelin, 1791

Beaded Periwinkle

This shell is found on rocks well above the high tide line and receives its moisture from salt spray. The outside of the shell is covered with small, pearly beads.

Beaded Periwinkle *Tectarius muricatus* Linné, 1758

CORAL REEFS

Few shells are found on live coral. They much prefer dead coral, coral rubble, and coral sand, all of which are found in abundance around the live reefs. Coral rubble is a particularly rich environment, since pieces of dead coral of various sizes are piled together with enough space to provide sanctuary for many types of shells. Shells are also found in the crevices around and under coral reefs. The divers shown above are holding large Cushion Stars which are common not only around reefs, but throughout the Keys.

Carrier Shell

If a shell collector turned into a seashell, he would probably be a carrier shell. This mollusk is the most famous collector in the ocean. Although other creatures, such as sea urchins, cover themselves with shells as camouflage, they can easily drop them. The live carrier shell uses a special cement to permanently affix its collection.

It is not clear why carriers collect other shells. Even carriers that live in the deep, totally dark depths of the ocean are known to collect, so the purpose may not be camouflage. One possibility is that the collected shells, which extend outward like the legs of a lunar landing craft, may help the carrier to avoid sinking into mud.

So how does the carrier reach up to affix collectibles to the very top of its shell? The answer is simple. The first shell at the top was put in position when the carrier was small and immature. Later, as its own shell grew wider, it added row after row of shells around the bottom.

Atlantic Carrier Shell *Xenophora conchyliophora* Born, 1780

Atlantic Carrier Shell

Carrier Shells (not from Florida)

◁ Carriers in different parts of the world decorate their shells with strikingly different material including rocks, sponges, and live coral.

83

Atlantic Thorny Oyster

Thorny Oyster

These shells are often attached to rocks and underwater objects, including sunken ships, but may also live unattached on sandy bottoms. They are not really oysters despite their common name but are related to the scallops. When taken live they are often covered with heavy growths of sponges and algae. Because of the spectacular beauty of their long spines, they are called Chrysanthemum Shells.

◁ A cluster of three thorny oysters showing the wide variety of colors.

Atlantic Thorny Oyster
Spondylus americanus
Hermann, 1781

Frog Shell

This uncommon shell is distinguished by the delicate color around its opening, or aperture.

The St. Thomas Frog Shell was first discovered in St. Thomas, Virgin Islands, hence the name *thomae*. It is found in Southeast Florida and up the West Coast to the Panhandle. It lives under rocks and coral rubble from very shallow water to water that is several hundred feet deep.

◁ The St. Thomas Frog Shell feeds on marine worms and small bivalves.

St. Thomas Frog
Shell *Bursa thomae*
Orbigny, 1842

Lion's Paw
Scallops

Little
Knobby
Scallops

Deep-Water Scallops

The Lion's Paw is the largest American scallop and probably the most highly prized scallop in this country. It is distinguished by the large knobs on its ribs. These knobs are hollow and are filled with liquid in the live scallop. The function of such knobs and ribbing in various shell species is to strengthen the shell as a defense against predators.

Both the Lion's Paw and the Little Knobby Scallop are collected by diving or by dredging. The Little Knobby has knobs on its back but is more prized for the rich color of its interior.

Lion's Paw Scallop *Lyropecten nodosus* Linné, 1758
Little Knobby Scallop *Chlamys imbricatus* Gmelin, 1791

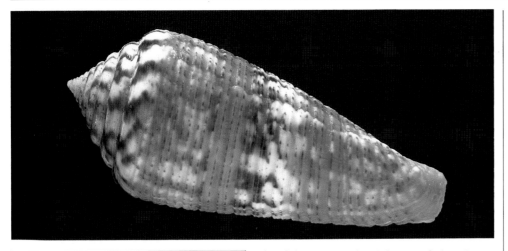

Glory-of-the-Atlantic Cone

This shell is now considered rarer than the famous Pacific cone. Glory-of-the-Seas. Even a small Glory-of-the-Atlantic is now often more valuable than the once highly prized Glory-of-the-Seas. Live specimens are usually obtained by scuba divers working at night with lights. Dead specimens are sometimes found in coral rubble along the Keys, but the lips are almost always broken. A perfect specimen is a true collector's item.

Glory of the Atlantic Cone *Conus granulatus* Linne, 1758

Florida Miter

This rare and beautiful shell can be found crawling in the sand around coral reefs. It is usually collected around reefs at night by divers or by dredging in deep water. Note the pointed shape, the spiral folds near the opening, and the many rows of orange dots.

Florida Miter *Mitra florida* Gould, 1856

File Clam

The File Clam is also called a Flame Scallop or a Lima. Its long tentacles break off easily and serve to entangle and discourage predators. The shells of this creature are rather plain and do not excite collectors, but the live animal is one of the most beautiful in the sea.

File Clams tend to congregate in groups and, although they can swim, they prefer the sedentary life, attaching themselves in crevices with their byssal threads. The shells are rather small, less than two inches, but appear much larger with the live animal spilling out in every direction.

Spiny Lima Lima lima *Linné, 1758*

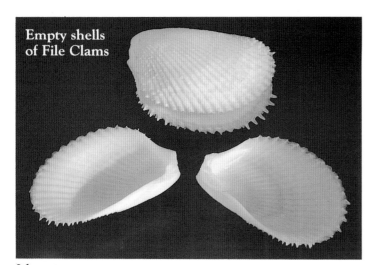

Empty shells of File Clams

Tritons

Triton's Trumpet

The Triton's Trumpet is a highly-prized shell. It can sometimes be found in shallow water and around rocks and bridges as well as around coral reefs. It eats starfish. This shell can reach a length of more than 16 inches. Notice the "teeth" on the outer lip which are grouped together in pairs.

This shell is named for the Greek demigod Triton, son of Neptune, whose blowing on a shell horn causes the roaring of the seas. The tip of the shell is filed off and it is played like a bugle in many parts of the world.

Atlantic Hairy Triton (with and without periostracum)

Dog-head Tritons

△ The Hairy Triton can be identified by the "teeth" around the shell opening. The live Hairy Triton grows a thick, outer coating (periostracum) which helps protect the shell from boring sponges. It is one of the few shells found in both the Atlantic and Pacific Oceans.

The live animal is beautifully colored and looks like a little giraffe. It is usually found on rocks and coral reefs in the Keys but is occasionally found farther north along both coasts (*see also Angular Triton, page 69*).

Triton's Trumpet *Charonia variegata* Lamarck, 1816 • Atlantic Hairy Triton *Cymatium pileare* Linné, 1758 • Dog-head Triton *Cymatium moritinctum* Clench & Turner, 1957

OTHER SHORE LIFE IN THE KEYS

Decorator Crabs

Crabs

Decorator Crabs

△ Several species of crab are called Decorator Crabs because they conceal themselves with sponges and vegetation. The crab will hold a piece of sponge against its shell until it begins to grow there. Soon it will be totally covered with sponge.

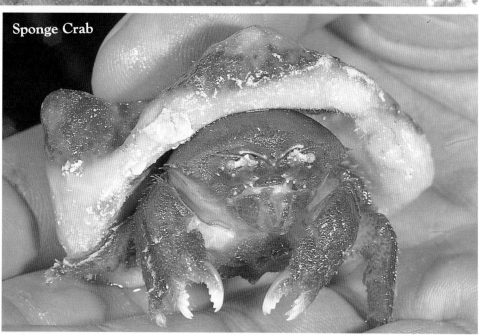

Sponge Crab

Arrow Crab

Coral Crab

◁ The Coral Crab is one of Florida's most beautiful Crabs. It occurs among coral reefs and rocky rubble, and it is highly prized as food in the Caribbean islands.

Land Crab

Great Land Crab

▷ This large crab lives in burrows on the shore under vegetation. Although shy, if you remain still and quiet it will emerge from its burrow and prowl around in plain view. It has one enormous claw. It can climb trees in its search for food. It is itself highly regarded as food on the Caribbean islands.

A well-camouflaged Purse Crab

Great Land Crab

Crabs

Land Hermit Crabs

Land hermit crabs are very colorful and can be purchased in pet shops and kept in terrariums as pets. They will eat cat food but thrive on nuts and fruit. Hermit crabs delight their owners with their beautiful colors and curious expressions. They do, however, have powerful claws and can give a painful pinch if handled carelessly.

Hermit crabs, like the horseshoe crabs, are not true crabs and, unlike the true crabs, do not move sideways. Out of their shells they somewhat resemble lobsters.

Hermit Crab "Conventions"

Land hermit crabs are known to gather at certain times of the year in groups as large as several hundred. They meet at a place where there are some empty shells. They will look over each others' shells and the empty shells and then do some trading. One crab will step out of its shell, and another will race over to claim the empty home.

△ Notice in the photo of the hermit crab out of its shell that its tail is curved so as to fit easily into a spiral shell.

Hermit Crab in shell of a Crown Conch

Hard-Tube Worms

Christmas Tree Worm retracted in its tube

Emerging from the tube

Tentacles fully extended for feeding

There are many different kinds of beautiful worms that live in hard tubes attached to rocks under water. The part of the worm that emerges from the tube looks very much like a feather duster and is composed of many feathery tentacles that gather small food particles from the water. One of the tentacles of this type of worm is modified to plug the tube, like the operculum of a shell, when the worm retreats into its home. These worms are easily seen by looking into the clear water of the Keys along rocky seawalls.

Feather Duster Worm fully extended

Coral polyps and Feather Duster Worm (not fully extended).

Hard-tube worms are all very sensitive to changes in light intensity. If a curious scuba diver swims past and casts his shadow upon them, they will pop back into their tubes.

The frilly parts that extend from the tube, are not venomous and are not harmful to the human touch. Other marine worms which are not attached in a tube, should be handled with caution. Some such as the Bristle Worm (which is common in the Keys), can inflict a painful sting.

Sea Anemones

Flat Sea Anemone

There are many different species of sea anemones found in the Keys. Many can be seen when walking along the rocky shore, and even more are visible to the snorkeler.

Sea anemones often attach themselves to the shells of marine hermit crabs and ride around with them so that they can collect small particles of food which drop in the water as the hermit crabs tear into their meals. In some instances the hermit crab may actually select an anemone and place it upon its shell.

Ringed Sea Anemone

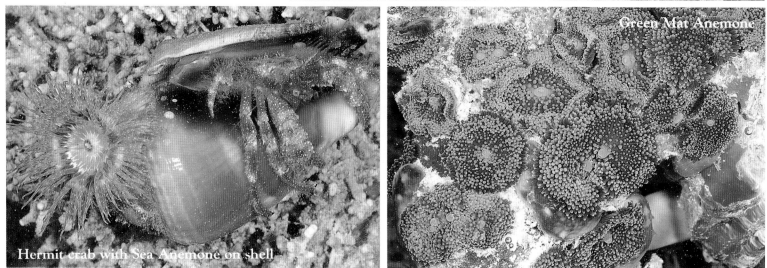

Hermit crab with Sea Anemone on shell

Green Mat Anemone

Anemone Shrimp on Pink-tipped Anemone

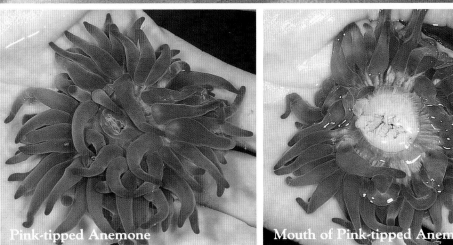
Pink-tipped Anemone

Mouth of Pink-tipped Anemone

The Spiny Lobster of the Keys is highly prized for food. It can obtain a length of over 2 feet and weigh more than 30 pounds.

"Bang, You're Dead"

This colorful creature is called the Pistol Shrimp because of its unique hunting technique. One claw is greatly enlarged and one finger of this claw fits snugly into a receptacle on the opposing finger. The finger is cocked open by a special mechanism and then released with great force and a loud underwater noise, something like a cap pistol. This noise from the "snapping" of the fingers is sufficient to stun a smaller fish or another shrimp for a second or two, just enough time for the Pistol Shrimp to harvest its meal. If the Pistol Shrimp is in an aquarium, the noise of its pop gun is loud enough to be heard in another room.

Purple Pistol Shrimp

Sea Slugs

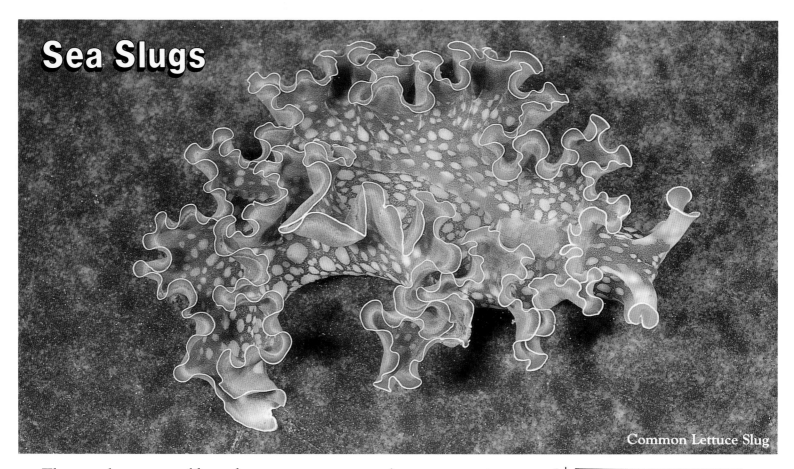

Common Lettuce Slug

The sea slugs, or nudibranchs, are marine snails without shells. They are among the most beautifully colored and lavishly decorated creatures in the sea. There are many species, each displaying a different and wildly spectacular palette of colors. Most are small, less than an inch in length, and most have a pair of antennae. They are found in shallow waters around sponges, tunicates, and anemones.

Nudibranch means "naked gills," and refers to the frilly skin on the back. Nudibranchs breathe through their skin, and the loose folds of skin increase the surface area over which gases are exchanged.

Basket Stars

These magnificent creatures ordinarily look like messy balls of twine, all tangled and twisted. This specimen has been stretched out to show its remarkable shape and placed in a garbage can lid to show its size. At its center, the basket star shows the same five armed radial symmetry as other starfish, but each arm quickly subdivides into many complicated shapes. Basket stars are usually found curled up on sea whips in 20 to 40 feet of water.

Rock-boring Urchin

Brittle Stars

Urchins

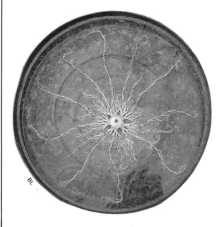

Brittle stars or serpent stars have long, slender, flexible legs that allow them to move much faster than their starfish relatives. Like starfish, they can regenerate lost body parts. There are many different species in Florida waters, especially in the Keys.

Both the Long-spined Urchin and the Rock-boring Urchin are threats to swimmers. Their spines contain barbs and toxins and tend to break off beneath the flesh. The Rock-boring Urchin is common in shallow water, the Long-spined around reefs.

Sand Dollars of the KEYS

Arrowhead

Sea Biscuit

Pancake

Florida has four varieties of sand dollars. They are called Flat Sand Dollars, Arrowhead Sand Dollars, Sea Biscuits, and Pancake Sand Dollars. The last three found mostly in the Florida Keys, although Sea Biscuits are abundant in Northwest Florida, and Arrowheads are sometimes found along the West Coast. The Pancake Sand Dollar is jumbo-sized and lacks the poinsettia-shaped markings on the bottom, as does the Sea Biscuit (*see page 28 for details about the Flat Sand Dollar*).

Portuguese Man-of-War

This is one of the most dangerous creatures in Florida waters. It is not common on the West Coast but fairly common along the lower Atlantic Coast and in the Florida Keys. It has long, stinging tentacles which may trail through the water for up to sixty feet on the largest specimens. The creature can extend the tentacles for fishing and retract them to harvest the catch. These stingers can cause extreme pain and even death to sensitive individuals. Even after the animal has beached itself and appears dead, the stingers may still be dangerous. Dive boats in the Florida Keys frequently carry a jar of meat tenderizer on board as an antidote for the sting.

The inflatable pouch with a "sail" on top allows the Portuguese Man-of-War to move through the water when driven by the wind and current, but it cannot steer and is at the mercy of these forces. It is the triangular shape of this "sail" that reminded English sailors of the slanting sails of a Portuguese vessel and inspired the name.

The Portuguese Man-of-War is not a single organism but a colony of many organisms, all living under the single float. Remarkably, each of the organisms contributes a different function such as producing the float, bearing the stingers, or digesting the food.

95

DEEP WATER SHELLS

Zigzag Scallops

Ravenel's Scallops

Zigzag Scallop

The Zigzag Scallop gets its name from the zigzag markings on its upper lid. These markings are quite distinct on some specimens and completely absent on others. Part of the appeal of this shell is the variety of its patterns and colors.

Notice that the upper half of the shell is almost flat, while the lower half is quite concave, like a cup. Sometimes the top valves of the Zigzag Scallop, or the very similar Ravenel's Scallop, wash up on Florida beaches. Tourists call them "fan shells," and many who find them are not aware that they have found only half of the shell of a bivalve.

Ravenel's Scallop has a flat upper valve like the Zigzag but lacks the distinctive zigzag markings. Ravenel's Scallop also has more prominent ribs on both the upper and lower valves.

Zigzag Scallop *Pecten ziczac* Linné, 1758
Ravenel's Scallop *Pecten raveneli* Dall, 1898

Sundial

This shell was once called the Architecture Shell because of its beautiful spiraling. The scientific name for this worldwide genus is *Architectonica*. According to legend, Leonardo Da Vinci designed a spiral staircase inspired by this shell.

Common Sundial *Architectonica nobilis* Röding, 1798

Junonia

These shells are the pride of Sanibel Island. Once in a while they wash up on Sanibel beaches after storms and are a truly rare find for the sheller wishing to add to his assortment of "self-collected" shells. They are hauled in from deep water by shrimpers in fair quantities, so there is no shortage of them in the shell shops. The Junonia, unlike most other snails, has no "trapdoor" (operculum) covering the shell opening.

Junonia *Scaphella junonia* Lamarck, 1804

Live Junonia

Spirula

These small coiled shells which resemble rams' horns are the internal supports for a small deepsea squid. When the squid dies and its flesh rots away, the shell floats to the surface of the ocean and is washed ashore.

The inch-wide shell is composed of many gas-filled chambers, in the manner of the famous Chambered Nautilus. Notice the pearly, concave door to the last chamber which is visible at the shell opening.

Common Spirula *Spirula spirula* Linné, 1758

Flame Auger

This is one of the most sought-after Florida shells and one of the largest augers in the world (5 inches or more). It has become quite scarce in recent years as its population has come under pressure from collectors.

Flame Auger *Terebra taurinus* Lightfoot, 1786

Bonnets

Many of these fine shells are tossed onto Florida beaches after storms. Bonnets have a wide "shield," a flat area adjoining the shell opening (but not as wide as the helmet shells). Their puffed up, bulging shape is said to resemble a lady's bonnet. The live bonnet lays its eggs in a tall, tower-like column. The bonnet is the state shell of North Carolina.

Bonnets feed on sand dollars and sea urchins. They bore holes in the hard outer coverings and extract the edible material.

Scotch Bonnet *Phalium granulatum* Born, 1778

Paper Nautilus

The Paper Nautilus is a parchment-like structure secreted by an argonaut, a creature that is very similar to the octopus. This enclosure serves as a brood chamber in which eggs are laid and also provides shelter for the female argonaut which curls up inside it. However, the argonaut is not attached to the Paper Nautilus and may abandon it at

Scotch Bonnets

Paper Nautilus

will. The argonaut floats near the surface in warm waters all over the world. The egg cases occasionally wash up onto Florida beaches. These cases can be rather small or over a foot in length, depending upon the species.

Cabrit's Murex

This uncommon shell is found on sand or coral rubble in deep water. It is Florida's answer to the Venus Comb Murex, a famous shell of the Pacific Ocean which many people consider the most beautiful murex in the world (*see page 106 for comparison*).

A group of Cabrit's Murexes cooperates by laying their eggs together in a single mass which may be a foot high. This is in contrast to most other kinds of shells which generally lay their eggs individually and attach the mass to some support.

The shells are obtained from scallop dredgers, so they do appear in shell shops, but it is always difficult to obtain a good specimen with perfect spines.

Cabrit's Murex *Murex cabritii* Bernardi, 1859

Distorsios

The name "distorsio" relates to the distorted shape of the shell. This Florida shell is similar to its famous Pacific relative, *Distorsio anus*. A view of the opening of this shell will give a clue to the origin of this scientific name.

Atlantic Distorsio *Distorsio clathrata* Lamarck, 1816

Cowrie-helmet

The Reticulated Cowrie-helmet is sometimes found in shallow water, more often in deeper water around coral reefs. It looks like a cross between a helmet shell and a cowrie.

Reticulated Cowrie-helmet *Cypraecassis testiculus* Linné, 1758

Atlantic Distorsio

How Rare Shells Are Collected in the Philippines

Many of the very rare, deep-water shells found in the Philippines are too fragile for dredging, so the harvesters drop very fine, weighted tangle nets onto the ocean floor just before sunset. The shell creatures which come out of the sand to crawl around at night get tangled in these nets and are pulled up in the morning. This method is becoming less common because of the enormous amount of labor required to keep these delicate nets in good repair.

REALLY DEEP WATER

Beau's Murex

This shell is very difficult to obtain and is a prized collector's item. It lives in very deep water (more than 600 feet) off Southwest Florida and must be dredged. Only a few specimens have turned up recently, so at the moment it can only be obtained from old collections. A feature of this shell is the beautiful webbing between the spines.

Beau's Murex *Murex beauii* Fischer & Bernardi, 1857

The Scallop Scavengers

Florida shellers used to go to Cape Canaveral to visit a dump where commercial scallopers disposed of all the unwanted shells and marine life dredged up along with their scallops. This was an important source of deep water shells that ordinarily do not reach the beach in good condition. Helmets, cones, bonnets, Egg Cockles, olives, doves, the big tun shells, cowries, and Giant Atlantic Murex were found. Shellers even cut open the little Bat Fish to find the tiny shells that the fish had eaten. These dumps are no longer open to the public. Scallop processors have become aware of the value of these deep water shells and now pull out the valuable ones to sell to collectors. Also, the collectors made a nuisance of

Shellers at scallop dump

themselves, sometimes even jumping onto trucks as they were dumping their loads. One woman was buried under a load of shells and had to be rescued with a crane. Still, even today, resourceful shellers have their contacts among the commercial scallop fishermen.

Scallop dredging is accomplished by dragging a metal-edged basket across the sea bottom. This dredge is indiscriminate and simply picks up everything in its path. There is considerable destruction of all kinds of marine life.

100

Superb Gaza

This beautiful top shell is distinguished by its iridescent green color. It was once a very rare find but now is more easily obtained from shrimpers trawling for the Royal Red Shrimp in deep waters of the Northern Gulf of Mexico. Dealers and collectors often put black cotton into the Gaza to make the shell appear greener. The shell at right in the top photo is filled with black cotton. Notice the difference in color.

Superb Gaza *Gaza superba* Dall, 1881

The Best Shelling Spot in Florida

According to one expert, the location of this secret place is somewhere near the Florida/Georgia border. When tourists from the north head home with a load of live shells which they collected in South Florida and neglected to clean, this is the point at which the smell becomes unbearable. They pull into the wayside park and head for the dumpster which is already overflowing with bags of shells from other tourists, making it the richest accumulation of seashells in the state.

Beach Access — The Law

In Florida all beaches may be used by the public up to the mean high water mark, and sometimes even further where the public has gained an easement by actually using the property over many years. The mean high water mark is usually determined by the piles of seaweed cast the farthest up the beach, but there are frequent disputes about the proper location for the "No Trespassing" signs of private land owners. Local governments can make regulations for the use of beaches such as curfews, bans on alcoholic beverages, limits on motor vehicles, etc. The real problem however, is getting to the beach. Access may be blocked by private property. This issue has become a complex legal and political problem. Beach replenishment programs (pumping more sand onto eroded beaches) which are financed with state money now require that adequate beach access be provided for the public.

△ These beautiful fossil shells are just a small part of what was found by one man in one afternoon's exploration of a mining pit

FOSSIL SHELLS

A Window to an Ancient World

Few people outside of shelling circles are aware of the fascinating fossil shells which are readily available to anyone interested. Excavations of ancient shell material are going on throughout the state. The material that is mined in these huge open pits is very, very old and contains many shell species which are no longer in existence.

There are several types of fossil deposits in Florida ranging in age from 100,000 to 50 million years. The entire peninsula was under the ocean in ages past, so deposits of shells can be discovered in almost every part of the state.

These ancient shell fossils contain chemical material which is similar to that of modern shells. They are not "mineralized" like certain other kinds of fossils (such as plants and animals) in which the original material has been replaced. These shells are called "fossils" because they are very old.

Fossil Nutmeg

Modern Nutmeg

△ Compare the ancient fossil nutmeg shell with the modern version and note the changes that have occurred over millions of years.

Ecphora

◁ One of the spectacular fossil species, *Ecphora bradleyae* Petuch, 1987, which has no modern equivalent.

Shelling 20 miles from the beaches

Mining operations are conducted in the open pits with drag lines and other excavation equipment to reach the shell-rich layers 10 to 20 feet beneath the surface. The shell material is then washed to remove the sand and the shells are sorted into several sizes and sold as foundation material for roads and parking lots.

A paleontologist (person who studies fossils) who has been working the pit near Sarasota every week for the last five years says that during each visit he finds something he has never seen before. As the machines dig deeper, new layers are uncovered. Often small pockets are opened which contain quantities of a certain shell found nowhere else in the pit.

Fossil shells can be found many places outside of the mining pits. One reason is that ancient rivers have washed a lot of this material toward the sea and deposited it in various places, including the beaches. Another reason is that during the building boom, fill dirt containing fossil shells was used for road and building foundations.

Most of Florida's fossil deposits are on private property where mining operations are in progress. The owners of these open pit mines are usually not very happy about having the public roaming around. By the late 1990s, all the commercial shell mining operations had been closed to the public.

Shellers at the fossil pit.

Equipment for sorting shell fill material

Looking into the strata

Fossilized Lion's Paw

Burying Knowledge

The scientifically invaluable material dug from these pits is not bound for any museum (except what is salvaged by collectors). Most of it is used for fill for new roads and parking lots. Countless ancient life forms never seen by man are pulled from the ground unnoticed, transported, and buried again as someone constructs a new driveway.

103

Sharks' teeth

Why Only the Teeth of Sharks?

A shark's teeth are the only part of the shark's skeleton containing calcium carbonate, a hard mineral which survives long enough to become fossilized. The rest of the shark's skeleton is not made of bone like that of other animals but is composed of organic cartilage which rots immediately upon the creature's death.

The fossilized teeth of many sea creatures can be found on the beach. But sharks' teeth are more numerous because sharks lose teeth and regenerate new ones throughout their lives. A large shark may produce 20,000 teeth during its lifetime. Also, sharks' teeth are more obvious because they are much larger and less likely to have been worn down through the years.

The giant tooth of a Carcharodon Megalodon, predecessor to the Great White Shark. This creature has been extinct for thousands of years which is fortunate for us since it may have been 50 to 100 feet in length. Its huge teeth are sometimes found by snorkelers and divers off the beaches near Venice, Florida. Four-inch specimens are not unusual and the largest reach eight inches.

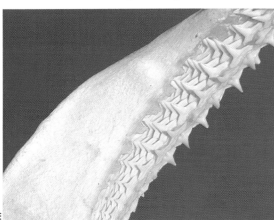

△ This view of a shark's jaw shows rows of new replacement teeth lined up, waiting to advance forward as old teeth are lost.

▷ Included in this photo is a piece of fossil bone which is frequently found along with sharks' teeth and the mouth plate of a ray. The serrated side of the plate is the root side, and the flat side is the surface the ray uses to crush crabs and shells.

Fossil Bone

Mouth Plate of Ray

Fossil Sharks' Teeth

Beach Sand

Black sand beach at Venice, Florida

Although sharks' teeth can be found on every beach in Florida, the best location is Venice Beach, from the Venice jetty to the south end of Casperson Beach. The fossil deposits in this area are so heavy that the sand is very dark, almost black. A visit to this beach will certainly result in the collection of many sharks' teeth.

Fossil oyster shells on black beach sand

The Color of Beach Sand

Beachcombers who travel around Florida will notice that the sand on Florida beaches is different in color and consistency from place to place. Here are the most common types of sand.

Pure white sand: The Siesta Key community in Sarasota claims the whitest sand in the world. It is also found at Ft. Walton Beach, Panama City Beach and other places in the Florida Panhandle.

This type of sand is made from quartz and gypsum. Quartz is formed volcanically, but Florida has never had volcanic activity.

All of the white sand now on Florida beaches has washed down from the Appalachian Mountains where it was originally formed. It has not yet arrived in the Keys and it never made it to the Bahamas because of the fast-moving Gulf Stream current.

Deposits of pure white sand are also found inland along ancient coastlines. Highway 27 south of Lake Wales follows a ridge of this material which was once an ancient seashore. Deposits of white sand can be seen at Oscar Scherer State Recreation Area south of Sarasota, Highlands Hammock State Park near Sebring, and along the west side

of the Indian River on the East Coast. Certain plants, such as the Scrub Oaks, thrive in this dry sand.

Black sand: Venice Beach is known for its black sand which is made of ground-up fossils such as sharks' teeth, bone, etc. This sand, although quite dark when wet, is not as black as the sand of beaches in other parts of the world that are made from volcanic rock.

Brown sand: Beach sand on St. George Island in Northwest Florida, Turtle Beach on Siesta Key, Sarasota, and in the Keys is made mainly from crushed shell material.

SHELLING AS A HOBBY

Cleaning Shells

Entire books have been written about cleaning shells and many species have individual requirements, but here are the basics: Remove the live creature from its shell by freezing or boiling and using a fork or a pair of pliers to twist the meat out of the shell. Soak the shell overnight in bleach to remove any remaining scraps of meat and also to remove algae and crud from the exterior. Use a plastic or glass container because bleach will react with metal. (Do not bleach olives or cowries, as the bleach will dull their shiny finish.) Use a knife or dental pick to scrape away any hard encrustations. Give the shell a

final cleaning by brushing it with a stiff toothbrush using warm, soapy water, and then allow it to dry. Some collectors like to coat their specimens with a small amount of a mixture of 2/3 baby oil and 1/3 lighter fluid. This brings out the shell colors and protects the outer surface of the shell to prevent it from drying and losing more color. Other collectors think that this gives an unnatural appearance and prefer to omit this final step. Shells used for crafts or decoration are frequently given a clear glaze for extra strength and protection.

How They Do It in the South Pacific

In the Pacific islands shells are cleaned by throwing them under a palm tree and allowing the ants to eat the fleshy parts. Often, shells are shipped to American importers in burlap bags in a rather raw and smelly condition. The importers finish the cleaning job before selling the shells to local shell shops.

△ Transparent "boat" for viewing bottom and storing specimens in shallow water.

Florida Snow Shovels

◁ These scoops are sold in local hardware and beach stores. They are handy for sifting shells and take the back-breaking labor out of collecting.

Famous Shells Not Found in Florida

The shells in this photo are frequently seen on display in shell shops. But don't expect to find these shells in Florida. Some of the most spectacular shells can only be found in other parts of the world. Here are a few examples (from top left, clockwise): Chambered Nautilus, Papal Miter, Fluted Giant Clam, Heart Cockle, Glory-of-the-Seas Cone, Japanese Wonder Shell, Map Cowrie, and Venus Comb Murex (center).

How Shells Are Judged at Shows

Shell shows are an opportunity for collectors to get together and display their collections. For those of a competitive nature, prizes are awarded in different divisions.

There are many categories at shell shows. Some collectors specialize in varieties of just one species or one family. Others collect miniatures, freak shells, shells of a certain locality, shells of a particular color, fossil shells, shells of a certain beach, etc. The theme for a collection is limited only by a sheller's imagination.

Florida has the largest share of shell shows. They are sponsored by the local shell clubs and are held mostly during the winter months.

The Tiny Shells

Among shell collectors, there is a specialty called "miniatures" which includes all adult shells less than one inch in length. In this group, a 1/2 inch shell is a giant. Surprisingly, there are more species of miniature shells than there are of larger shells. Many shells are so small they can only be seen with a magnifying glass. Collectors find them by sifting through sand at high magnification. Also, some collectors seek the tiny, baby specimens of large shells.

The largest of these shells is 1/4 inch.
▽

Find 'em or Buy 'em

At shell shows there is a division between collectors who find their own shells (self-collected shells) and collectors who obtain their shells from any source (shell dealers, trading, trips abroad, etc.). It might seem that the latter category would be simply a competition in spending money between wealthy collectors, but this is not true in practice. Factors other than money are important to the judges. A collection is rated on its theme, its attractiveness and its instructional value, as well as the rarity of the shells displayed.

What's So Special About the Operculum?

The operculum or "perk" is the "trapdoor" to the opening of a marine snail's shell. The word itself means lid. It is attached to and part of the soft body of the mollusk, and looks something like a hard, leathery callus. The live animal uses this piece of hard material to close the entrance after it has retracted back into its spiral shell.

Some Non-Florida Turbans

Many shellers like to include the operculum with their shells because they feel it shows the specimen more completely. Other people find this practice aesthetically distasteful. Virtually all shell books state somewhere that *serious* collectors always *save the operculum,* implying that those who do not are hopeless amateurs. Usually the operculum is glued to a piece of cotton which is stuffed into the shell.

Most opercula are rather dark and lacking in appeal, but the opercula of some shells, particularly the turbans, have their own beautiful color and patterns. The color and shape of opercula can aid in the identification of some shells.

◁ The colorful opercula of some non-Florida shells.

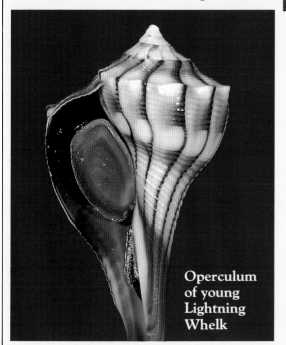

Operculum of young Lightning Whelk

△ This Lightning Whelk has closed its trapdoor for protection.

Shellcraft

Most shell shows have a division for shell craft competition, and there are many categories, including flowers, critters, lamps, mirror frames, mosaics, miniatures, and traditional shell valentines.

The Story of the Sailor's Valentine

The traditional Sailor's Valentine must have the following features: flowers, a heart shape, and a verse. A Sailor's Valentine is usually two-sided and hinged so that it will close (usually with a heart-shaped lock). Victorian style valentines are one-sided.

Sailor's Valentines supposedly were the work of homesick sailors at sea, but the Barbados newspapers found glued on the backings and the generic verses, such as "Love the Giver," seem evidence that they were not made for a special loved one by some devoted sailor but were purchased from craft people as an afterthought before the ship reached its final port.

Sailor's Valentine

Collecting the Shells for Shell Art

Many shell crafters buy their shells from dealers, but others take great pride in collecting everything they use in their work. Crafters also take pride in using only sea materials. They might construct the leaves of a bouquet as well as the flowers from shells. Sea materials other than shells are often employed. Crafters may, for instance, cut Sand Dollars into certain shapes and combine them with shells.

△ Shell crafters can be rather competitive.

◁ Shell crafters use tiny shells to create miniatures. These shells are not easy to obtain. Formerly, dealers dug buckets of tiny shells from the surf and paid nursing home residents to sort them. The most valued shells were resold per thimbleful.

OTHER SHELL WORLDS

Florida Tree Snail

These snails are of considerable interest to collectors because of their great beauty and variety. Of the hundreds of species of land snails in the United States, only this South Florida species, *Liguus*, produces beautifully colored shells. *"Liguus"* means "banded." Some believe that the brightly banded patterns featured in Seminole Indian fabrics were inspired by the color bands of these snails.

Land development has eliminated most of the South Florida habitat suitable for *Liguus* except for certain areas within the Everglades National Park. *Liguus* Tree Snails are now mostly found on Everglades hammocks (small hardwood forests on raised land surrounded by wet habitat). *Liguus* Tree Snails boast a wide range of colors and their isolation on the island-like hammocks has helped the many color forms maintain their purity.

Florida Tree Snail *Liguus fasciatus* Müller, 1777

Liguus Tree Snails live on smooth-barked trees where they feed on lichens. They mow a swath through the lichens growing on the bark, leaving a visible path on the trunks of the trees where they are feeding. They are active during spring and summer, but seal their shells and hibernate on branches during the dry winter months.

Fresh Water Mollusks

These shells are less colorful than sea shells and less popular with collectors (except, perhaps, collectors in land-locked parts of the country). However, the interiors of certain freshwater clams contain some of the most beautiful mother-of-pearl.

There are far fewer species of fresh water mollusks than of their salt water relatives, and their shells are generally more fragile because the fresh water environment is less turbulent.

Buckley's Pearly Mussel *Elliptio buckleyi* Lea, 1853

SKIES AND BEACHES

Waterspouts

A waterspout is a tornado which forms over the water. Generally, the waterspouts seen in Florida are less powerful than tornados in the midwest but they put on a spectacular show. It is not unusual to see several funnel clouds at one time. If a waterspout comes ashore, then it is called a tornado and can be very destructive. In Florida, they have been particularly damaging to trailer parks, although legislation has increased the requirements for tie-downs and other safety features in modular homes.

The Green Flash

The green flash occurs during sunset just as the last tiny fraction of the sun's sphere dips below the horizon. There is a bright flash of green color visible just above the ball of the sun, which lasts only a few seconds. Seeing the green flash may be slightly more difficult than finding a Junonia, but the rarity of the occasion makes it all the more satisfying.

Why does it occur? Even the disproven theories are interesting. It is not caused by the sun's yellow light passing through or reflecting off of blue sea water (it can occur over dry land). It is not the result of a mental image of a complementary color formed as a result of retinal fatigue from staring at the setting sun (it has been captured on photographic film).

The generally accepted theory is that some of the sun's rays passing through the atmosphere at a very low angle are broken into the colors of the spectrum in the same manner as light passing through a prism. The prism colors are not visible before the sun sets because of the intense light of the ball of the sun. The colors red, orange and yellow are beneath the sun and disappear when the sun sets. The colors green, blue, indigo, and violet are above the ball of the sun. At the instant that the sun dips below the horizon, there is a darkening of the sky and the remaining spectrum colors above the sun momentarily become visible. Blue, indigo, and violet are more susceptible to scattering in the atmosphere, so it is usually only the green that is visible, although a blue flash is sometimes seen seconds before sunrise.

Since haze increases scattering of light, the green flash is more likely to be visible when the air is clear and clean. Binoculars can also help because the band of color is actually very small. However, to avoid eye damage, the binoculars should not be pointed at the sun until just before the last edge of the disk disappears.

The well known novelist and Sarasota resident, John D. MacDonald, entitled one of his books *The Flash of Green* in honor of this phenomenon (most of his books include the name of a color in their titles). Another author, inspired by the flash in a very different era was Jules Verne, who wrote a novel called *The Green Ray*.

Sunlight is broken into the colors of the prism by the atmosphere. Prism colors are not visible at this time because the sun is too bright.

Sky darkens when sun sets. Blue and violet prism colors are scattered in the atmosphere but the green is momentarily visible in the darkened sky.

Red, orange and yellow prism colors have disappeared below the horizon.

TABLE OF CONTENTS

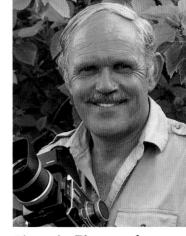

About the Photographer

Many nature magazine and book editors regard Pete Carmichael as the world's finest shell photographer. Although best known for his photos of beach life and butterflies, Pete's work covers a broad range of nature subjects and has been published in all the major nature magazines such as National Geographic, Audubon, National Wildlife, and Natural History. Pete has a master's degree in anthropology and has taught advanced photography courses for over 15 years.

All the photos in this book are from Pete's vast collection or selected by him from the best shots of his students and shelling friends.

Acknowledgements:

The author is grateful to the following consultants for their generous help with this project. R. Tucker Abbott, Ph.D., Norman Blake, Ph.D., Al and Bev Deynzer, Pete Carmichael, Mack Hamby, Ed Hanley, Dan and Brenda Hanley (Sunray Shell Charters), Skip Hengen, Charles and Vi Hertwick, Russ Jensen, Bob and Betty Lipe, Bill Lyons, John Morrill, Ph.D., Larry Rabinowitz, John Stevely, Don Thomas, Bill Tiffany, Ph.D., Sue Vaughan, and Richard Walden.

Photo Credits:

All photos are by Pete Carmichael except those photos where the photographer's initials appear beside the photo according to the following key:

David Addison/Visuals Unlimited (DA/VU), Larry Andrews (LA), Stephen Frink/Water House (SF/WH), Patricia Hanbery (PH), Bob Lipe (BL), Bob Pelham (BP), Gail Shumway (GS), Lynn Stone (LS), Bill Tiffany (BT), Milton Tierney/Visuals Unlimited (MT/VU) Joanne Werwie (JW), Winston Williams (WW), and Sondra Williamson (SW)

Cover credits: Front cover: Pink Conch (GS); inside front cover: Lion's Paw Scallop (GS); Page 1: Shellers in Tampa Bay (BL); Inside Back Cover: Pete Carmichael; Back Cover: (clockwise from top left) Lion's Paw, Horse Conch, Christmas Tree Worm, Girl with Conch, Thorny Oyster, Hermit Crab, Imperial Venus Clam (center), all by Pete Carmichael.

Cartoons by Brian Yoong

Florida's Fabulous Seashells and Seashore Life
by Winston Williams
© 2009 World Publications
11th Revised Edition
L.C. No. 86-051449   ISBN: 0-911977-05-8
Manufactured in Singapore
For book orders, please contact:
World Publications
Post Office Box 1850
Hawaiian Gardens, CA 90716
Phone: 562-924-8300
Fax: 562-924-8305

Other Titles